Future Trends in Modern Plastics

Scrivener Publishing
100 Cummings Center, Suite 541J
Beverly, MA 01915-6106

Publishers at Scrivener
Martin Scrivener (martin@scrivenerpublishing.com)
Phillip Carmical (pcarmical@scrivenerpublishing.com)

Future Trends in Modern Plastics

Johannes Karl Fink
Montanuniversität Leoben, Austria

Scrivener Publishing

WILEY

This edition first published 2024 by John Wiley & Sons, Inc., 111 River Street, Hoboken, NJ 07030, USA and Scrivener Publishing LLC, 100 Cummings Center, Suite 541J, Beverly, MA 01915, USA

For more information about Scrivener publications please visit www.scrivenerpublishing.com.

Wiley Global Headquarters
111 River Street, Hoboken, NJ 07030, USA

For details of our global editorial offices, customer services, and more information about Wiley products visit us at www.wiley.com.

Library of Congress Cataloging-in-Publication Data

ISBN 9781394237548

Cover image: Pixabay.com
Cover design by Russell Richardson

Set in size of 11pt and Minion Pro by Manila Typesetting Company, Makati, Philippines

Printed in the USA

10 9 8 7 6 5 4 3 2 1

Contents

Preface

This book focuses on the issues concerning future trends in plastics.

The book begins with a chapter about monomers and polymerization methods. Here, newly developed monomers, such as alkylene-based monomers, epoxide monomers, diol-based monomers, bio-based monomers, and several other types, are discussed. Then, modern polymerization methods are explained, such as ionic polymerization, plasma polymerization, and ring-opening polymerization.

Then, in the next chapter, special issues and some future trends in the plastics industry are explained. Here, recommendations for future research, are also noted. Also discussed are the enormous benefits plastics have brought to society owing to their versatility, light weight, durability and low costs. However, these properties have come with negative externalities, especially because these persistent materials are leaked into the environment during their entire life cycle. Therefore, an important section is included on the future directions for sustainable polymers.

The valorization of plastic waste is an important feature of Chapter 4. Nowadays, polymers are the most versatile materials. They contain certain chemicals and additives, such as pigments, concentrates, anti-blockers, light transformers, UV-stabilizers, etc. Therefore, an in-depth analysis has been presented with respect to the recovery, treatment and recycling routes of plastic waste in Chapters 5 and 6.

The methods of characterization are detailed in Chapter 7. Here, the properties and material testing methods, such as standards, are described.

In Chapter 8, usage of plastics in medical devices are detailed. Here, the properties, requirements, and applications are presented along with a comprehensive overview of the main types of plastics used in medical device applications.

The subsequent Chapters and their subject matter are the use of plastics in restoration, food applications, additive classes and manufacturing.

The text focuses on the literature of the past decade. Beyond education, this book will serve the needs of specialists who have only a passing knowledge of the subject matter but need to know more.

How to Use This Book

Utmost care has been taken to present reliable data. Because of the vast variety of material presented here, however, the text cannot be complete in all aspects, and it is recommended that the reader study the original literature for more complete information.

The reader should be aware that mostly US patents have been cited where available, but not the corresponding equivalent patents in other countries. For this reason, the author cannot assume responsibility for the completeness, validity or consequences of the use of the material presented herein. Every attempt has been made to identify trademarks; however, there were some that the author was unable to locate.

Index

There are three indices: an index of acronyms, an index of chemicals, and a general index.

In the index of chemicals, compounds that occur extensively, e.g., "acetone," are not included at every occurrence, but rather when they appear in an important context.

Acknowledgements

I am indebted to our university librarians, Dr. Christian Hasenhüttl, Friedrich Scheer, Christian Slamenik, and Elisabeth Groß for their support in literature acquisition. Also, many thanks to Boryana Rashkova for her nice support.

I also want to express my gratitude to all the scientists who have carefully published their results concerning the topics dealt with herein. This book could not have been otherwise compiled.

Last, but not least, I want to thank the publisher, Martin Scrivener, for his abiding interest and help in the preparation of the text. In addition, my thanks go to Jean Markovic, who made the final copyedit with utmost care.

Johannes Fink
Wiener Neustadt
December 20, 2023

1

Monomers and Polymerization Methods

Several monomers are used for polymers. Most of them are old but some of them are fresh materials. Here, monomer types and monomers are given and also special methods for polymerization.

A lot of these materials are collected in books (1–6).

Monomers can be subdivided into two classes, depending on the kind of polymer that they form (7). Monomers that participate in condensation polymerization have a different stoichiometry than monomers that participate in addition polymerization. Classifications may also include (8):

1. Alkylene monomers
2. Epoxide monomers
3. Diols
4. Diacids
5. Amino acids
6. Alcohol acids
7. Bio-based monomers
8. Nucleotides
9. Monosaccharides
10. Natural monomers
11. Synthetic monomers
12. Polar monomers
13. Nonpolar monomers

1.1 Types of Monomers and Synthesis Methods

In this section, common monomers, both conventional and modern monomers, are shown.

1.1.1 *Alkylene Monomers*

Various monomer types are presented in Tables 1.1, 1.2, and 1.7 below. Also, these compounds are shown in Figures 1.1 and 1.2.

Table 1.1 Monomers with one double bond.

Compound	Compound
Ethylene	Propylene
1-Butene	1-Pentene
2-Butene	2,3-Dimethyl-1-butene
1-Pentene	2-Pentene
2-Methyl-1-butene	3-Methyl-1-butene
2-Methyl-2-butene	1-Hexene
2-Hexene	3-Hexene
2-Methyl-1-pentene	3-Methyl-1-pentene
4-methyl-1-pentene	2-Methyl-2-pentene
3-Methyl-2-pentene	4-Methyl-2-pentene
2,3-Dimethyl-1-butene	3,3-Dimethyl-1-butene
2,3-Dimethyl-2-butene	2-Ethyl-1-butene
α-Pinene	6,6-Dimethylbicyclo[3.1.1]hept-2-ene

Table 1.2 Monomers with multiple double bonds.

Compound	Compound	Compound
Butadiene	Isoprene	Chloroprene
Norbornadiene	1,5-Cyclooctadiene	Dicyclopentadiene

Some modern alkene-based monomers are shown in Table 1.3.

1.1.1.1 *Apopinene*

Apopinene (6,6-Dimethylbicyclo[3.1.1]hept-2-ene), c.f. Figure 1.4, is a biorenewable monomer that can be used for ring-opening metathesis polymerization (9).

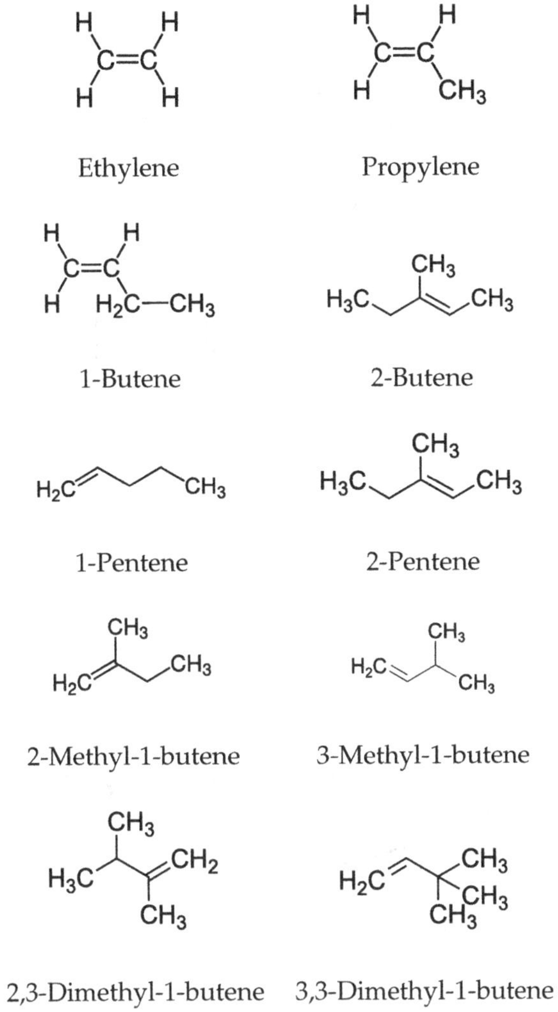

Figure 1.1 Monomers with one double bond.

Table 1.3 Modern Monomers.

Compound	Reference
Apopinene	(9)
6,6-Dimethylbicyclo[3.1.1]hept-2-ene	(9)
Bio-based acrylic monomers	(10)

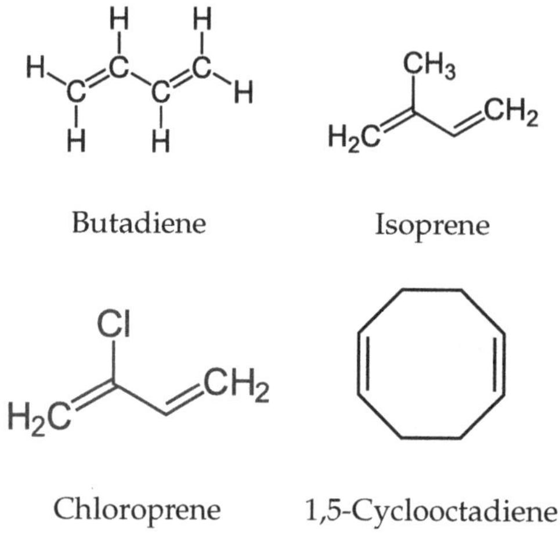

Figure 1.2 Monomers with multiple double bonds.

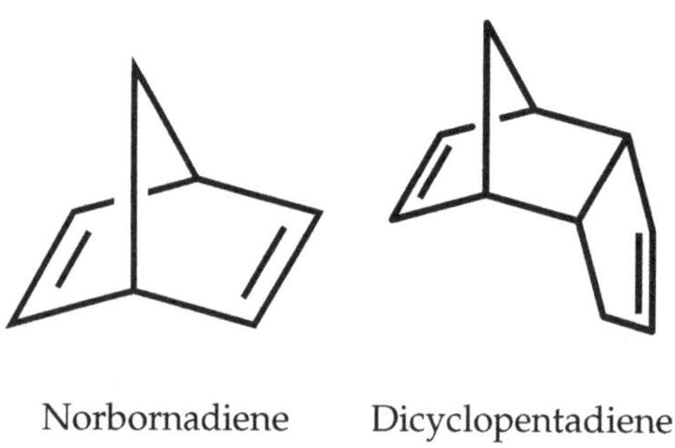

Figure 1.3 Cyclic monomers with multiple double bonds.

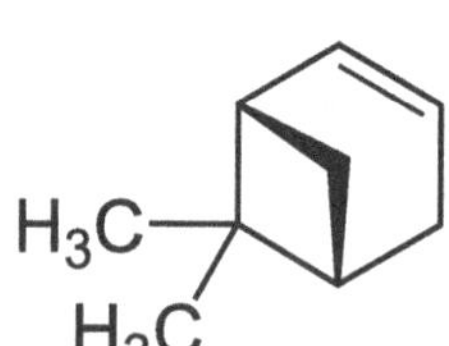

Figure 1.4 Apopinene.

Apopinene is the most abundant monoterpene present in nature and plays a crucial role in many biological, atmospheric and industrial processes. Similar to many other readily accessed and biorenewable terpenes, α-pinene is widely used in both the fine chemical and polymer industries. The Lewis acid-catalyzed polymerization of α-pinene generates a polymer and has found a variety of uses in a plethora of industrial applications such as adhesives, plastics, and rubbers.

The high abundance, low cost, and biorenewability of α-pinene make its incorporation into additional novel materials highly desirable from the standpoint of sustainability.

One avenue that has sparked some theoretical interest is the ring-opening metathesis polymerization of α-pinene (11).

1.1.2 Epoxide Monomers

Various epoxide monomers are presented in Table 1.4. Some of these monomers are also shown in Figure 1.5.

The synthesis of functionalized polycarbonates, employing mainly propylene oxide and cyclohexene oxide, has been detailed (12). In recent years, functionalized polycarbonates have become an emerging topic with a broad scope of potential applications. The synthetic routes and properties of numerous functionalized polycarbonates synthesized from CO_2 and functional epoxide monomers have been described (12).

The synthesis of polymers from renewable resources is of high interest. Polymeric epoxide networks constitute a major class of thermosetting polymers and are extensively used as coatings, electronic materials, and adhesives (13). Owing to their outstanding mechanical and electrical properties, chemical resistance, adhesion, and minimal shrinkage after curing, they are used in structural applications as well.

Most of these thermoset types are industrially manufactured from bisphenol A (BPA), a substance that was initially synthesized as a chemical estrogen (13). The awareness of BPA toxicity combined with the limited availability and volatile cost of fossil resources and the non-recyclability of thermosets implies necessary changes in the field of epoxy networks. Thus, substitution of BPA has witnessed an increasing number of studies both from the academic and industrial

Table 1.4 Epoxide Monomers.

Compound	Reference
Epoxy crotyl sucrose	(14)
Ethylene oxide	(15)
Propylene oxide	(15)
1,2-Butylene oxide	(15)
2,3-Butylene oxide	(15)
2,3-Epoxy heptane	(15)
Nonene oxide	(15)
5-Butyl-3,4-epoxyoctane	(15)
1,2-Epoxy dodecane	(15)
1,2-Epoxy hexadecane	(15)
1,2-Epoxy octadecane	(15)
5-Benzyl-2,3-epoxy heptane	(15)
4-Cyclo-hexyl-2,3-epoxy pentane	(15)
Chlorostyrene oxide	(15)
Styrene oxide	(15)
o-Ethylstyrene oxide	(15)
m-Ethylstyrene oxide	(15)
p-Ethylstyrene oxide	(15)
Glycidyl benzene	(15)
7-Oxabicyclo[4.1.0]heptane	(15)
Oxabicyclo[3.1.0]hexane	(15)
4-Propyl-7-oxabicyclo[4.1.0]heptane	(15)
3-Amyl-6-oxabicyclo[3.1.0]hexane	(15)

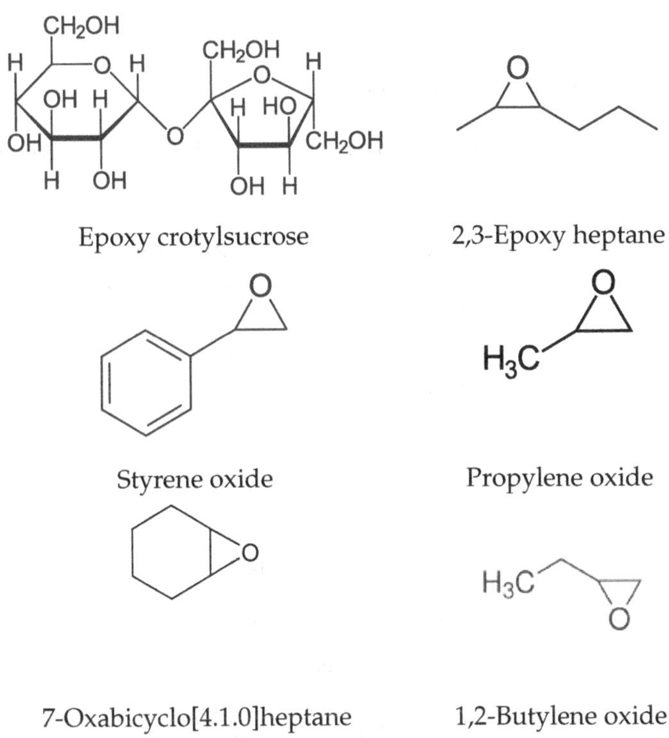

Figure 1.5 Epoxide monomers.

sides. This review presents an overview of the reported aromatic multifunctional epoxide building blocks synthesized from biomass or from molecules that could be obtained from transformed biomass.

The main glycidylation routes and mechanisms and the BPA toxicity were described. Also, the main natural sources of aromatic molecules have been detailed. The various epoxy prepolymers can be organized into simple mono-aromatic di-epoxy, mono-aromatic poly-epoxy, and derivatives with numerous aromatic rings and epoxy groups (13).

1.1.3 *Diol-Based Monomers*

Diol-based monomers are presented in Table 1.5 and shown in Figure 1.6.

Table 1.5 Diol based monomers.

Compound	Compound
1,6-Hexanediol	1,8-Octanediol
1,10-Decanediol	

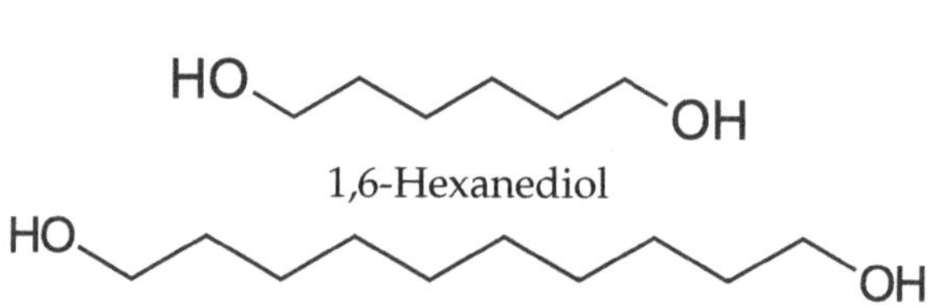

Figure 1.6 Diol-based monomers.

The synthesis and characterization of variants of poly(diol fumarate) and poly(diol fumarate-*co*-succinate) were described. Through a Fischer esterification, α,ω-diols and dicarboxylic acids were polymerized to form aliphatic polyester comacromers. Because of the carbon-carbon double bond of fumaric acid, incorporating it into the macromer backbone structure resulted in unsaturated chains.

By choosing α,ω-diols of different lengths (1,6-hexanediol, 1,8-octanediol, and 1,10-decanediol) and controlling the amount of fumaric acid in the dicarboxylic acid monomer feed (33, 50, and 100 mol%), nine diol-based macromer variants were synthesized and characterized for molecular weight, number of unsaturated bonds per chain, and thermal properties.

Degradation and *in vitro* cytotoxicity were also measured in a subset of macromers.

Macromer networks were photocrosslinked to demonstrate the ability to perform free radical addition using the unsaturated macromer backbone. Crosslinked macromer networks were also characterized for physicochemical properties (swelling, sol fraction, compressive modulus) based on diol length and amount of unsaturated bonds. A statistical model was built using data generated from these diol-based macromers and macromer networks to evaluate the impact of monomer inputs on final macromer and macromer network properties. With the ability to be modified by free radical addition, biodegradable unsaturated polyesters serve as important macromers in the design of devices such as drug delivery vehicles and tissue scaffolds. Given the ability to extensively control final macromer properties based on monomer input parameters, poly(diol fumarate) and poly(diol fumarate-*co*-succinate) represent an exciting new class of macromers (16).

1.1.4 Diacid-Based Monomers

Diacid-based monomers are shown in Table 1.6 and in Figure 1.7.

Table 1.6 Diacid based monomers.

Compound	Compound
2,5-Furan dicarboxylic acid	3,4-Furan dicarboxylic acid
2,3-Furan dicarboxylic acid	Adipic acid
Azelaic acid	

1.1.5 Bio-based monomers

Bio-based platforms for polymers are shown in Table 1.8.

Also, the routes to some representative sustainable polymers that are synthesized from biomass feedstocks were

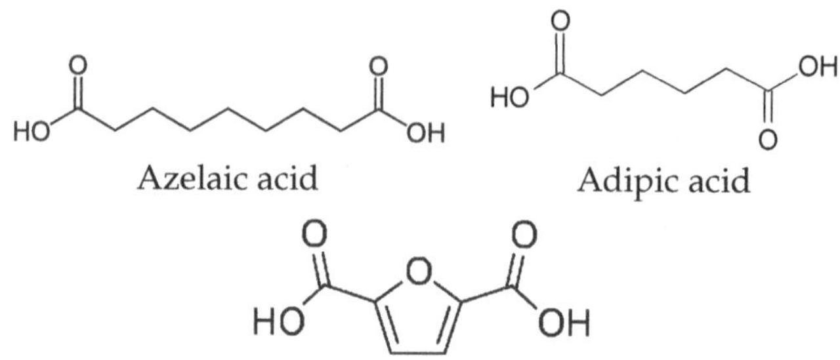

2,5-Furan dicarboxylic acid

Figure 1.7 Diacid based monomers.

Table 1.7 Bio-based monomers (17).

Compound	Compound	Compound
Hydroxy acids	Diacids	Diols
Diamines	Triglycerides	Fatty acids
Amino acids		

Table 1.8 Bio-Based Platforms (17).

Platform	Monomer type	Polymer type
Sugar	Hydroxy acids, diacids, diols, diamines, cyclics, vinyl	Copolyesters, polyester polyols, copolyamides, poly(urethane)s, polyolefins, polyacids
Lignin	Acids, alcohols	Polyesters, polybenzoxazines
CO_2	Cyclic carbonates Polycarbonates, non-isocyanate poly(urethane)s	
Vegetable oils	Triglycerides, fatty acids	Polyesters, poly(urethane)s, thermosets
Proteins	Amino acids, (macro)cyclics	Poly(amino acid)s, poly(ester urea)s, polydepsipeptides, poly(ester amide ester)s, peptoids, cationic polymers

shown (17). Natural biopolymers, including polysaccharides, lignin, lipids, polypeptides and terpenes, can be extracted from renewable biomass. Through deconstruction and conversion or fermentation, natural biopolymers can be turned into polymer precursors for sustainable polymerization. Sustainable polymers are poly(ethylene), poly(propylene), poly(ethylene 2,5-furandicarboxylate), poly(furfural alcohol), poly(hydroxyalkanoate)s, poly(lactide) and poly(butylene succinate).

The recent advances in the microbial production of diamines, aminocarboxylic acids, and diacids as potential platform chemicals and bio-based polyamides monomers have been described (18).

Bio-based manufacturing processes of chemicals and polymers in biorefineries using renewable resources have extensively been developed for the sustainable carbon dioxide (CO_2) neutral-based industry. Bio-based diamines, aminocarboxylic acids, and diacids have been used as monomers for the synthesis of polyamides with different carbon numbers and ubiquitous and versatile industrial polymers and also as precursors for further chemical and biological processes to afford valuable chemicals.

These platform biochemicals have been successfully produced by biorefinery processes employing enzymes and/or microbial host strains as main catalysts (18).

Metabolic engineering strategies of microbial consortia and optimization of microbial conversion processes, including whole cell bioconversion and direct fermentative production, have been developed.

1.1.6 Fatty Acids

Monomeric unsaturated fatty acids, which are derived from natural sources, are capable of being polymerized to the dimerized and trimerized form (19). This is usually realized by heating such unsaturated fatty acids in the presence of catalytic proportions of a mineral clay and, preferably, an acid-treated mineral clay, at temperatures in excess of about 180°C in an aqueous environment under autogenous pressure. Small amounts of water are deemed necessary for reaction to minimize the degradation of the fatty acids being treated (19).

Fatty acid monomers can be employed as reactive diluents for the polymerization of vinyl esters and polyesters (20, 21). They can improve the fracture resistance, lower the processing viscosity and reduce the volatile organic compounds that are present in the polymerization mixture.

Fatty acid monomers can be used to replace some or all of the styrene used in liquid thermosetting resins. They are excellent alternatives to styrene because of their low cost and low volatility.

Furthermore, fatty acids are derived from plant oils, and are therefore a renewable resource. Thus, not only would the use of fatty acids in liquid molding resins reduce health and environmental risks, but it also promotes global sustainability. Fatty acids and triglycerides have been used in a number of polymeric applications (20, 21).

The preparation of epoxidized and hydroxylated fatty acids has been reviewed (22–24).

Fatty acids that may be employed to synthesize fatty acid monomers are listed in Table 1.9. Some of these compounds are shown in Figure 1.8

Table 1.9 Fatty acids for fatty acid monomers.

Compound	Compound	Compound
Butyric acid	Capric acid	Caprylic acid
Lauric acid	Myristic acid	Palmitic acid
Stearic acid	Oleic acid	Linoleic acid

A polymeric composition, wherein the fatty acid monomer is a monomer of the formula, is shown in Figure 1.9.

Here, R is selected from the group consisting of a C_2−−C_{30} saturated alkyl residue, an unsaturated alkyl residue, an acetylenic alkyl residue, a hydroxyl alkyl residue, a carboxylic acid alkyl residue, a divinyl ether alkyl residue, a sulfur-containing alkyl residue, an amide alkyl residue, a methoxy alkyl residue, a keto alkyl residue, a halogenated alkyl residue, a branched methoxy alkyl residue, a branched hydroxyl alkyl residue, an epoxy alkyl residue, a fatty acyl-CoA alkyl residue, a cyclopropane alkyl residue, a cyclopentenyl

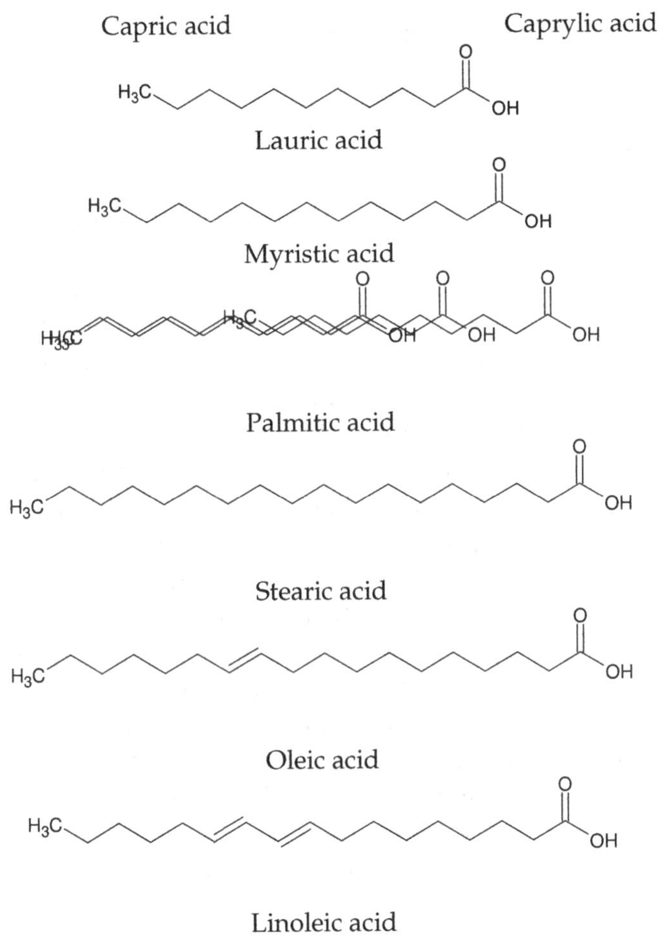

Figure 1.8 Fatty acids.

alkyl residue, a cyclohexyl alkyl residue, a furanoid alkyl residue, a phenylalkanoic alkyl residue, and a lipoic alkyl residue (20, 21).

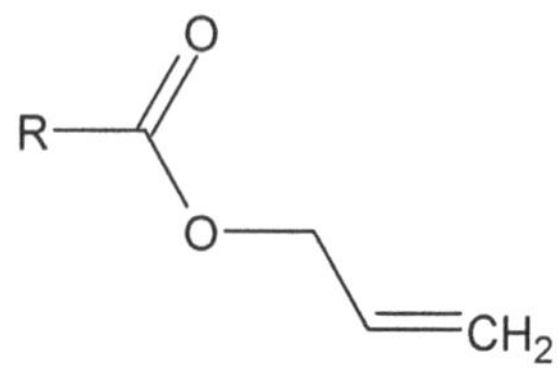

Figure 1.9 Fatty acid monomer.

1.1.7 Cyclic Fatty Acids

Cyclic fatty acid monomers have been found in frying oils used for fast foods (25).

Cyclic fatty acids can be classified into those that are naturally occurring and those that are formed in vegetable oils during heating (26). The former include cyclopropane, cyclopropene and cyclopentenyl acids. C17 and C19 cyclopropane acids are common in many bacteria, such as lactobacilli and enterobacteria; and mycolic (2-alkyl-3-hydroxy) acids, with up to about 90 carbons and one or two cyclopropane rings, occur in mycobacteria (27, 28).

Several acids with up to 26 carbons, one or two cyclopropane rings and a double bond in the 5 position were identified in an invertebrate from a deep-water lake (29). The C18 and C19 cyclopropane acids occur in varying amounts in the seed oils of some species of a few plant families, including Malvaceae and Sterculiaceae (27, 30). The cyclopropene counterparts are more widespread in these families, and cyclopentenyl acids are present in the seed oils of the family Flacourtiaceae, notably the genus *Hydnocarpus* (30).

Fatty acids with six-membered (31) and seven-membered rings have been characterized from the thermoacidophilic bacterium, *Bacillus acidocaldarius*.

1.1.8 Triglycerides

Epoxidized and acrylated triglycerides have been used as plasticizers and toughening agents (20, 21). In fact, the largest non-food use of triglycerides is the use of epoxidized soybean and linseed oils as plasticizers in poly(vinyl chloride). Epoxidized triglycerides have also been studied for their use as toughening agents in epoxy polymers.

The potential of epoxidized palm oil as plasticizer for poly(lactic acid) (PLA) was compared with commercialized epoxidized soybean oil (32). The plasticizers were melt-compounded into PLA at 3%, 5%, 10%, and 15%. As the aim was for the blends to be characterized towards packaging appropriate for food products, they were hot-pressed into 0.3 *mm* sheets, which is the approximate thickness of clamshell packaging. Fourier transform infrared spectroscopy confirmed the compatibility of the plasticizers with PLA. At similar loadings, epoxidized palm oil was superior in reinforcing elongation at break, thermal, and barrier properties of PLA.

The ductility of PLA was notably improved to 50.0% with the addition of 3% of epoxidized palm oil. From differential scanning calorimetry, the increase in crystallinity and the shifts in enthalpy of fusions in all plasticized blends denoted facilitation of PLA to form thermally stable α-form crystals.

The addition of epoxidized palm oil enabled PLA to become highly impermeable to oxygen, which can extend its potential in packaging an extensive range of oxygen-sensitive food (32).

1.1.9 *Ester-Based Monomers*

Some modern ester-based monomers are shown in Table 1.10.

The free-radical polymerization of dialkyl methylene malonate monomers using heat, UV light and peroxide has been described (33–35). Here, the monomer was prepared using traditional methods, which results in a monomer with low purity. The polymers are prepared via bulk polymerization. One would therefore not expect to be able to control polymer properties such as molecular weight and molecular weight distribution.

The polymerization of 1,1-disubstituted alkene compounds using anionic polymerization processes which are useful in the bulk polymerization of 1,1-disubstituted alkene compounds and processes which can operate at or near ambient conditions have been reported. Anionic bulk polymerizations may be initiated using a wide range of initiators, and may even be initiated by contact with certain substrates. Other bulk polymerization reactions may be ini-

Table 1.10 Ester-based monomers (35).

Compound	Compound
Methylene malonate	Dibutyl methylene malonate
Diethyl methylene malonate	Dihexyl methylene malonate
Dimethyl methylene malonate	Dipentyl methylene malonate
Butyl ethyl methylene malonate	Ethyl hexyl methylene malonate
Diisopropyl methylene malonate	Ethyl pentyl methylene malonate
Hexyl methyl methylene malonate	Butyl methyl methylene malonate
Diethoxyethyl methylene malonate	Dimethoxyethyl methylene malonate
Menthyl methyl methylene malonate	Methyl pentyl methylene malonate
Methyl propyl methylene malonate	Fenchyl methyl methylene malonate
Ethoxyethyl ethyl methylene malonate	Di-*N*-propyl methylene malonate
Ethyl methoxyethyl methylene malonate	Ethoxyethyl methyl methylene malonate
2-Phenylpropyl ethyl methylene malonate	Methoxyethyl methyl methylene malonate
2-Phenyl-1-propanol ethyl methylene malonate	

tiated by UV light. However, the bulk polymerization may limit the ability to control the structure of the polymer molecules and/or to be able to easily handle the resulting polymer composition or product. These difficulties in bulk polymerization may be particularly pronounced when manufacturing large quantities of polymer, where heat transport issues may occur, especially when there may be shear heat generated by the flow of the high viscosity polymer and/or heat emitted due to the inherent exothermic nature of the polymerization.

Bulk polymerization of 1,1-disubstituted alkene compounds also present a challenge when attempting to control the structure of the polymer by including one or more comonomers. The high viscosity of the intermediate polymer may present difficulties in preparing a block copolymer, such as by sequential addition of a first monomer system followed by a second monomer system into a reaction vessel.

Other problems may arise when attempting to control the structure of a random copolymer, where the reaction rates of the different monomers differ so that the monomers are not uniformly distributed along the length of the polymer molecular.

For example, copolymers including one or more 1,1-disubstituted alkene compounds prepared by bulk polymerization are typically expected to have a generally blocky sequence distribution and/or result in polymer molecules having a broad distribution of monomer compositions. As used herein, a copolymer having a generally blocky sequence distribution of monomers may be characterized as having a blockiness index of about 0.7 or less, about 0.6 or less or about 0.5 or less, or about 0.4 or less (35).

Although solution polymerization processes have been employed in free-radical polymerization processes to better control the polymer architecture, such processes have not generally been employed for the anionic polymerization of 1,1-disubstituted alkenes. When a solution polymerization system is employed with anionic polymerization methods, sub-ambient temperatures (e.g., less than 10°C, less than 0°C, or less than -20°C) are typically required to control the reaction. As such, in solution polymerization systems it may be necessary to use a cooling system and/or insulation for achieving and/or maintaining such a low reaction temperature. Additional difficulties in the polymerization of 1,1-disubstituted alkene compounds arise from the possibility of the anionic group of the growing polymer reacting with an acid, thereby terminating the reaction. Therefore, using an acid in polymerizing 1,1-disubstituted alkene compounds using anionic polymerization should be avoided.

Prior attempts at anionic polymerization processes for 1,1-disubstituted alkene compounds generally have had the following drawbacks (35):

1. Requirement that the systems have low polymer concentrations,
2. Have lacked the reproducibility for controlling molecular weight distribution, or
3. have undesirable reactant by-products.

So, there is a need for polymerization methods, systems, and resulting polymer compositions or products that allow an improved control of one or more of the following properties of a polymer containing one or more 1,1-disubstituted alkene compounds (35):

1. The weight average molecular weight,
2. The number average molecular weight,
3. The polydispersity index,
4. The zero-shear viscosity of the polymer,
5. The viscosity of the polymer system.

There is also a need for a polymerization process, which can be scaled-up to a reactor of about 20 l or more, or having a throughput of about 10 $kg\,h^{-1}$ or more. There is also a need for processes that result in a solution containing the polymer. Such solutions may be useful for applications such as paints, coatings, finishes, polishes, and adhesives (35).

It has been found that a monomer including a 1,1-disubstituted alkene can be anionically polymerized using a solution polymerization process to controllably produce polymers (e.g., to produce polymers having controlled molecular weight and/or structure).

In this solution polymerization process, the monomers are diluted by a solvent and the monomer and solvent form a single continuous phase. During the polymerization process the resulting polymer may be soluble in the solvent, or may precipitate from the solvent (35).

An example of the preparation may run as follows (35):

Preparation 1–1: Fenchyl-methyl methylene malonate is polymerized in solution. The solvent is tetrahydrofuran. A round-bottom flask is charged with about 9.0 *g* of tetrahydrofuran and about 1.0 *g* of the fenchyl-methyl methylene malonate. The mixture is stirred with a magnetic stirrer for about 5 *min*. Tetramethyl guanidine (TMG) is then added to the flask to activate the polymerization reaction. The molar ratio of monomer (fenchyl-methyl methylene malonate) to activator (tetramethyl guanidine) is about 1000 (i.e., 1000:1). The polymerization reaction is continued for about 1 *h* at a temperature of about 23°C. The polymerization process is monitored by taking small aliquots of solution and quenching the reaction in the aliquot by adding an acid. After the 1 *h* polymerization, a molar excess of trifluoroacetic acid (TFA) is added to the flask to quench the polymerization reaction. An aliquot of the solution is taken and characterized by NMR spectroscopy. Another aliquot of the solution is analyzed by gel permeation chromatography to measure the molecular weight distribution. The solution is then precipitated in cold (0°C) methanol. The polymer precipitates as a white powder. The precipitated polymer is filtered, dried and then characterized using differential scanning calorimetry. NMR spectroscopy at the end of the reaction shows no measurable presence of residual monomer.

The GPC indicates that the polymer has a first peak in molecular weight at about 2000 and a second peak in molecular weight at about 60,000. The polymer has a polydispersity index of about 1.43. The glass transition temperature of the polymer is about 151°C. In the homopolymerization of fenchyl-methyl methylene malonate, by varying the reaction conditions, the purity of the monomer, the activator concentration and the reaction temperature, the molecular weight distribution of the polymer may be varied between about 1 to 8 and glass transition of the polymer may be increased to be as high as about 190°C (e.g., when weight average molecular weight is high).

Tetramethyl guanidine and trifluoroacetic acid are shown in Figure 1.10.

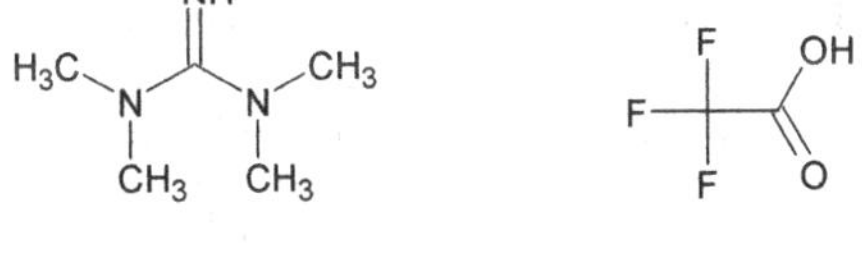

Tetramethyl guanidine   Trifluoroacetic acid

Figure 1.10 Tetramethyl guanidine and trifluoroacetic acid.

1.1.10 *Amino Acids*

Amino acids have emerged as a sustainable source for the design of functional polymers (36). Besides their wide availability, their high degree of biocompatibility makes them especially appealing for a broad range of applications in the biomedical research field. In addition to these favorable characteristics, the versatility of reactive functional groups in amino acids (i.e., carboxylic acids, amines, thiols, and hydroxyl groups) makes them suitable starting materials for various polymerization approaches, which include step- and chain-growth reactions (36).

Amino acids that can be used as monomers are presented in Table 1.11. Some of these compounds are also shown in Figure 1.11.

Table 1.11 Amino acids used as monomers.

Compound	Reference
8-Amino-2-quinolinecarboxylic acid	(37)
2,6-Diaminopyridine	(37)
2,6-Pyridinedicarboxylic acid	(37)
7-Amino-8-fluoro-2-quinolinecarboxylic acid	(37)
1,8-Diaza-9-fluoro-2,7-anthracenedicarboxylic acid	(37)
2-(4-Aminocyclohexyl)ethanoic acid	(38)
4(4-Aminocyclohexyl)butanoic acid	(38)
4-(Aminomethyl)cyclohexanecarboxylic acid	(38)
4-(2-Aminoethyl)cyclohexanecarboxylic acid	(38)

1.1.11 *Monosaccharides*

Monosaccharides are shown in Table 1.12. Some monosaccharides are also shown in Figure 1.12.

Table 1.12 Monosaccharides (39).

Compound	Compound	Compound	Compound	Compound
Allose	Altrose	Glucose	Mannose	Gulose
Idose	Galactose	Talose	Psicose	Fructose
Sorbose	Tagatose	Arabinose	Lyxose	Ribose
Xylose	Ribulose	Xylulose		

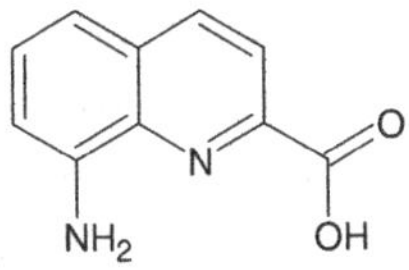

8-Amino-2-quinolinecarboxylic acid

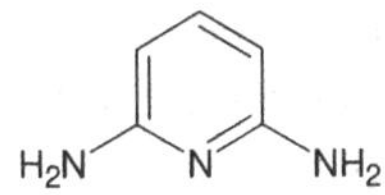

2,6-Diaminopyridine

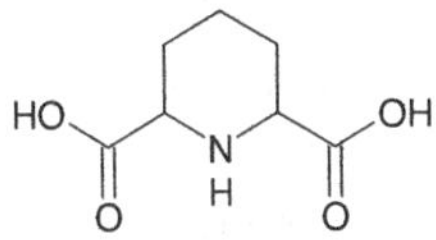

2,6-Pyridinedicarboxylic acid

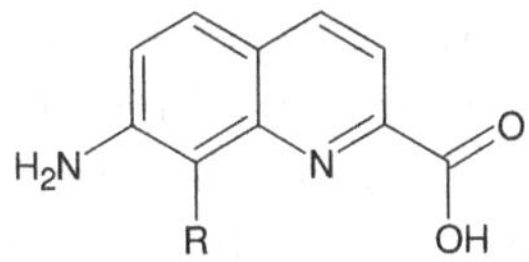

7-Amino-8-fluoro-2-quinolinecarboxylic acid

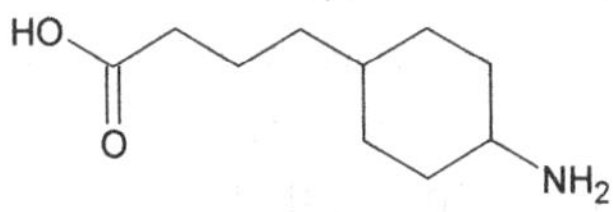

4(4-Aminocyclohexyl) butanoic acid

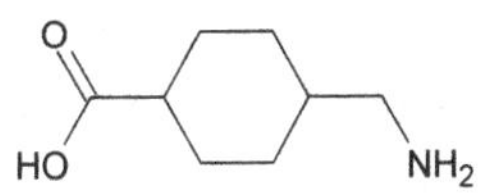

4-(Aminomethyl)cyclohexane-carboxylic acid

Figure 1.11 Amino acids used as monomers.

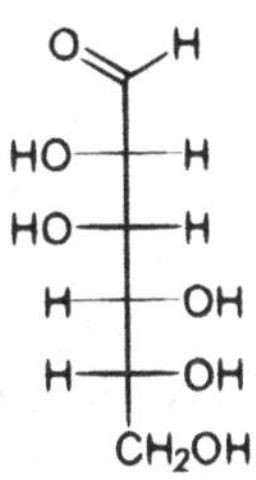

D-Mannose

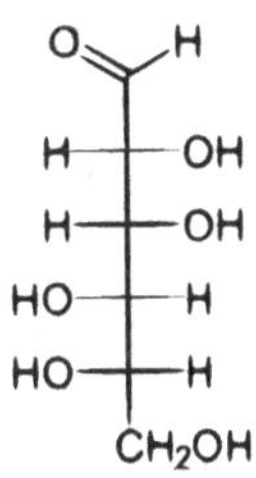

L-Mannose

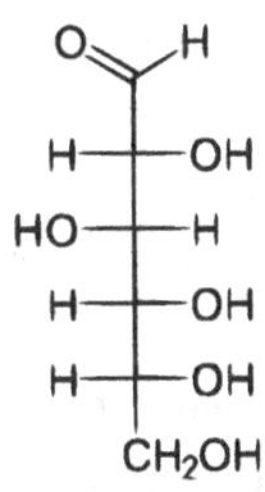

D-Glucose

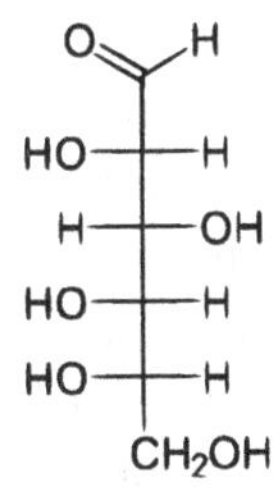

L-Glucose

Mannose

Glucose

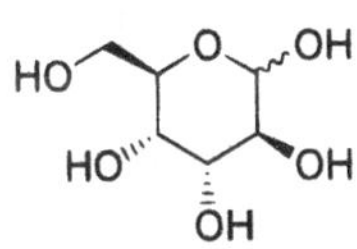

Altrose

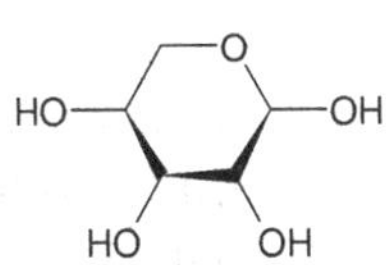

Lyxose

Figure 1.12 Monosaccharides.

A monosaccharide monomer may also be used as a mixture of several monosaccharide monomers. Furthermore, the copolymer can be prepared by using one or more diol monomers, such as octane-1,3-diol, *cis*-oct-5-ene-1,3-diol, isosorbide, 1,3-propanediol, 1,2-propanediol, and 1,4-butanediol (39).

Isosorbide is a bicyclic chemical compound from the group of diols and the oxygen-containing heterocycles containing two fused furan rings (40). The starting material for isosorbide is *D*-sorbitol, which is obtained by catalytic hydrogenation of *D*-glucose, which is in turn produced by hydrolysis of starch. Isosorbide is a plant-based platform chemical from which biodegradable derivatives of various functionality can be obtained.

A method for the production of monosaccharides from biomass has been described (41). In the first step, a raw material biomass is pretreated in 65 to 85% sulfuric acid at a temperature of 30°C to 70°C. In the second step, the product from the first step is subjected to a saccharification treatment in 20% to 60% sulfuric acid at a temperature of 40°C to 100°C. In detail, the method runs as follows (41):

First, 700 *g* of pine chips (coniferous tree) with a moisture content of 9.1% and containing 414 *g* of holocellulose and 1100 *g* of 71.5% sulfuric acid were charged in a mixing stirrer having a reactor volume of 10 *l*, followed by carrying out the pretreatment of the first step for 40 *min* at 50°C. The determination of the amount of sulfuric acid based on 100% conversion yielded a value of 786.5 *g*. Subsequently, hot water was charged in the reactor to dilute the sulfuric acid concentration to 30% followed by carrying out the saccharification treatment of the second step for 90 *min* at 85°C. At this time, in order to investigate the degree of degradation of xylose present in the saccharification treatment liquid (i.e., second step treatment product), the concentration of xylose was measured every 10 *min* using a high-performance liquid chromatograph. The relationship between saccharification treatment time and xylose concentration is shown in Table 1.13.

Based on the results of Table 1.13, since the xylose concentration was nearly constant, a reduction in the amount of xylose during saccharification treatment attributable to degradation was not observed. Next, the saccharification treatment liquid was cooled to about 40°C, and the filtration procedure of the second step was carried out.

Table 1.13 Saccharification treatment time and xylose concentration (41).

Reaction time *min*	Xylose concentration % by weight
10	0.30
20	0.38
30	0.43
40	0.45
50	0.48
60	0.51
70	0.51
80	0.52
90	0.50

The monosaccharide concentration of the resulting filtrate was measured using the aforementioned high-performance liquid chromatograph (HPLC). The amount of monosaccharide contained in the filtrate was calculated using the following equation based on that value and the total amount of liquid.

The amount of monosaccharides such as glucose, xylose and mannose, contained in the filtrate was 249 *g* after the saccharification treatment. The determination of the conversion rate of holocellulose to monosaccharides based on the weight of the holocellulose from the amount of monosaccharides yielded a value of 60.1%.

The concentration of sulfuric acid in this effluent saccharification liquid (raffinate) was 1.0% by weight. The monosaccharification treatment of the third step was then carried out on this effluent saccharification liquid (raffinate) using an autoclave by holding at a temperature of 121°C for 30 *min*.

Then, the sugar liquid was collected and the monosaccharide concentration in the sugar liquid was again measured using the aforementioned high-performance liquid chromatograph (HPLC) to calculate the amount of monosaccharides.

As a result, the amount of monosaccharides in the sugar liquid was 312 *g* (41).

1.1.11.1 Properties of Monosaccharides

The properties of monosaccharides can be found in detail in Wikipedia (42).

Allose. Allose is an aldohexose sugar. It is a rare monosaccharide that occurs as a 6-O-cinnamyl glycoside in the leaves of the African shrub *Protea rubropilosa*. Extracts from the fresh-water alga *Ochromas malhamensis* contain this sugar.

Altrose. Altrose is an aldohexose sugar. *D*-Altrose is an unnatural monosaccharide. *L*-altrose has been isolated from strains of the bacterium *Butyrivibrio fibrisolvens*.

Glucose. Glucose is a sugar with the molecular formula $C_6H_{12}O_6$. Glucose is the most abundant monosaccharide. It is mainly made by plants and most algae during photosynthesis from water and carbon dioxide using energy from sunlight, where it is used to make cellulose in cell walls. It is the most abundant carbohydrate in the world.

Mannose. Mannose is a sugar monomer of the aldohexose series of carbohydrates. It is a C-2 epimer of glucose. Mannose is important in human metabolism, in particular for the glycosylation of certain proteins.

Gulose. Gulose is an aldohexose sugar. It is a monosaccharide that is very rare in nature, but has been found in archaea, bacteria and eukaryotes. Also, it may act as a syrup with a sweet taste. It is soluble in water and slightly soluble in methanol. *d*-Gulose is a C-3 epimer of *d*-galactose and a C-5 epimer of *l*-mannose.

Idose. Idose is a hexose. It has an aldehyde group and is an aldose. It can be synthesized by the aldol condensation of *d*-glyceraldehyde and *l*-glyceraldehyde. *l*-Idose is a C-5 epimer of *d*-glucose.

Galactose. Galactose is a monosaccharide sugar that is about as sweet as glucose, and about 65% as sweet as sucrose. It is an aldohexose and a C-4 epimer of glucose. A galactose molecule linked with a glucose molecule forms a lactose molecule.

Talose. Talose s an aldohexose sugar. It is an unnatural monosaccharide. It is a C-2 epimer of galactose and a C-4 epimer of mannose.

Psicose. *d*-Psicose ($C_6H_{12}O_6$), also known as *d*-allulose, is a low-calorie epimer of the monosaccharide sugar fructose used by some major commercial food and beverage manufacturers as a low-calorie sweetener. It was first identified in wheat in the 1940s. Actually, allulose is naturally present in small quantities in certain foods.

Fructose. Fructose is a ketonic sugar found in many plants, where it is often bonded to glucose to form the disaccharide sucrose. It is one of the three dietary monosaccharides, along with glucose and galactose, that are absorbed by the gut directly into the blood of the portal vein during digestion. The liver then converts both fructose and galactose into glucose, so that dissolved glucose, known as blood sugar, is the only monosaccharide present in circulating blood.

Sorbose. Sorbose is a ketose. It has a sweetness that is equivalent to sucrose (table sugar). The commercial production of vitamin C (ascorbic acid) often begins with sorbose. *l*-Sorbose is the configuration of the naturally occurring sugar. It can be prepared from O-benzylglucose.

Tagatose. Tagatose is a hexose monosaccharide. It is found in small quantities in a variety of foods. Therefore, it has attracted attention as an alternative sweetening agent. It is often found in dairy products, because it is formed when milk is heated. It is similar in texture and appearance to sucrose. Since it is metabolized differently from sucrose, tagatose has a minimal effect on blood glucose and insulin levels. Tagatose is also approved as a tooth-friendly ingredient for dental products. The consumption of more than about 30 *g* of tagatose in a dose may cause gastric disturbances, as it is mostly processed in the large intestine, similar to soluble fiber.

Arabinose. Arabinose is an aldopentose, a monosaccharide containing five carbon atoms, and including an aldehyde functional group.

For biosynthetic reasons, most saccharides are almost always more abundant in nature as the *d*-form, or structurally analogous to *d*-glyceraldehyde. However, *l*-arabinose is in fact more common than *d*-arabinose in nature and is found in nature as a component of biopolymers such as hemicellulose and pectin.

Lyxose. Lyxose is an aldopentose, a monosaccharide containing five carbon atoms, and with an aldehyde functional group. Lyxose occurs only rarely in nature, for example, as a component of bacterial glycolipids.

Ribose. Ribose is a simple sugar and carbohydrate with molecular formula $C_5H_{10}O_5$. The naturally-occurring form, *d*-ribose, is a component of the ribonucleotides from which RNA is built.

Xylose. Xylose is a sugar first isolated from wood, and named for it. Xylose is classified as a monosaccharide of the aldopentose type, which means that it contains five carbon atoms and includes an aldehyde functional group. It is derived from hemicellulose, one of the main constituents of biomass. Like most sugars, it can adopt several structures depending on conditions. With its free aldehyde group, it is a reducing sugar.

Ribulose. Ribulose is a ketopentose, a monosaccharide containing five carbon atoms, including a ketone functional group. It has chemical formula $C_5H_{10}O_5$. Two enantiomers are possible, *d*-ribulose (*d*-erythro-pentulose) and *l*-ribulose (*l*-erythro-pentulose). *d*-Ribulose is the diastereomer of *d*-xylulose.

Ribulose sugars are composed in the pentose phosphate pathway from arabinose. They are important in the formation of many bioactive substances. As the 1,5-bisphosphate, *d*-ribulose combines with carbon dioxide at the start of the photosynthesis process in green plants (carbon dioxide trap). Ribulose has the same stereochemistry at carbons 3 and 4 as the five-carbon aldoses ribose and arabinose.

Xylulose. Xylulose is a ketopentose, a monosaccharide containing five carbon atoms, and including a ketone functional group. It has the chemical formula $C_5H_{10}O_5$. In nature, it occurs in both the *l*- and *d*-enantiomers. 1-Deoxyxylulose is a precursor to terpenes via the DOXP pathway.

1.1.12 Nucleotides

Nucleotides are organic molecules composed of a nitrogenous base, a pentose sugar and a phosphate (43). They serve as monomeric units of the nucleic acid polymers; deoxyribonucleic acid and ribonucleic acid. Both of these are essential biomolecules within all life forms on Earth. Nucleotides are obtained in the diet and are also synthesized from common nutrients by the liver.

Nucleotides are composed of three subunit molecules (43):

1. A nucleobase,
2. A five-carbon sugar (ribose or deoxyribose), and
3. A phosphate group consisting of one to three phosphates.

The four nucleobases in deoxyribonucleic acid are guanine, adenine, cytosine and thymine; in RNA, uracil is used in place of thymine. These compounds are shown in Figure 1.13

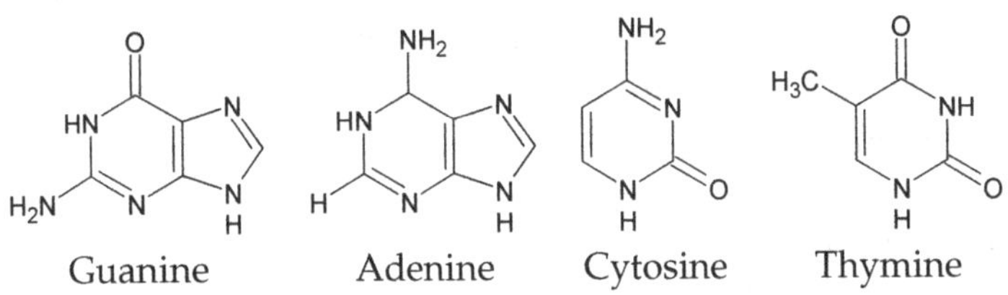

Figure 1.13 Nucleobases in DNA.

Nucleotides also play a central role in metabolism at a fundamental, cellular level. They provide chemical energy in the form of the nucleoside triphosphates, adenosine triphosphate (ATP), guanosine triphosphate (GTP), cytidine triphosphate (CTP) and uridine triphosphate (UTP) throughout the cell for the many cellular functions that demand energy, including amino acid, protein and cell membrane synthesis, moving the cell and cell parts (both internally and intercellularly), and cell division. In addition, nucleotides participate in cell signaling (cyclic guanosine monophosphate or cGMP and cyclic adenosine monophosphate or cAMP), and are incorporated into important cofactors of enzymatic reactions (e.g., coenzyme A, FAD, FMN, NAD, and NADP+) (43).

1.2 Polymerization Methods

1.2.1 *Anionic Polymerization*

Anionic polymerization is a form of chain-growth polymerization or addition polymerization that involves the polymerization of vinyl monomers initiated with anions. The type of reaction has many manifestations, but traditionally vinyl monomers are used. Often anionic polymerization involves living polymerization, which allows control of the final structure of the polymer.

1.2.2 *Cationic Polymerization*

Cationic polymerization is a type of chain-growth polymerization where a cationic initiator transfers an electric charge to a monomer which then becomes reactive.

This reactive monomer goes on to react similarly with other monomers to form a polymer. The types of monomers necessary for cationic polymerization are limited to alkenes with electron-donating substituents and heterocycles.

Similar to anionic polymerization reactions, cationic polymerization reactions are very sensitive to the type of solvent used. Specifically, the ability of a solvent to form free ions will dictate the reactivity of the propagating cationic chain.

1.2.3 *Plasma Polymerization*

Plasma polymerization (or glow discharge polymerization) uses plasma sources to generate a gas discharge that provides energy to activate or fragment gaseous or liquid monomer, often containing a vinyl group, in order to initiate polymerization. The polymers that are formed from this technique are generally highly branched and highly crosslinked, and adhere to solid surfaces well.

The biggest advantage to this process is that polymers can be directly attached to a desired surface while the chains are growing.

1.2.4 Ring-Opening Polymerization

Ring-opening polymerization is a form of chain-growth polymerization in which the terminus of a polymer chain attacks cyclic monomers to form a longer polymer. The reactive center can be radical, anionic or cationic.

Some cyclic monomers, such as norbornene or cyclooctadiene, can be polymerized to high molecular weight polymers by using metal catalysts. Ring-opening polymerization is a versatile method for the synthesis of biopolymers.

Ring-opening of cyclic monomers is often driven by the relief of bond-angle strain. Thus, as is the case for other types of polymerization, the enthalpy change in ring-opening is negative.

1.2.4.1 Monomers

Cyclic monomers that are amenable to ring-opening polymerization include epoxides, cyclic trisiloxanes, some lactones, lactides, cyclic carbonates, and amino acid *N*-carboxyanhydrides. Many strained cycloalkenes, such as norbornene, are suitable monomers via ring-opening metathesis polymerization.

1.3 Future Trends in Summary

This chapter explained monomers.

In addition, newly developed monomers, such as alkylene-based monomers, epoxide monomers, diol-based monomers, bio-based monomers, and several other types were discussed. Then, modern polymerization methods were explained, such as ionic polymerization, plasma polymerization, and ring-opening polymerization.

References

1. A. Ballada and G. Zaikov, *Preparation and Properties of Monomers, Polymers and Composite Materials.*
2. L. Bretschneider, *Orthogonal pathway design for eco-efficient biocatalytic monomer production from cyclohexane [Kumulative Dissertation].* PhD thesis, Dissertation, Halle (Saale), Martin-Luther-Universität Halle-Wittenberg, 2021, (2021).
3. R.K. Sadhir and R.M. Luck, *Expanding Monomers,* CRC Press Boca Raton, (1992).
4. A. D'Amore and G. Zaikov, *New Topics in Monomer and Polymer Research.*
5. J.W. Summers and G. Zaikov, *Basic Research in Polymer and Monomer Chemistry.*
6. K. Takemoto, R.M. Ottenbrite, and M. Kamachi, *Functional Monomers and Polymers,* CRC Press, (1997).
7. C.E. Carraher Jr, *Introduction to Polymer Chemistry,* CRC Press, (2012).
8. Wikipedia contributors, Monomers — Wikipedia, the free encyclopedia, https://en.wikipedia.org/wiki/Fourth_Industrial_Revolution, (2023). [Online; accessed 13-August-2023].
9. R.J. Thomson, B.F. Strick, F.M. Geiger, and M. Delferro, Apopinene as a biorenewable monomer for ring-opening metathesis polymerization, US Patent 9 850 340, assigned to Northwestern University, December 26, 2017.
10. A. Voronov and I. Tarnavchyk, Bio-based acrylic monomers and polymers thereof, US Patent 20 170 226 050, assigned to North Dakota State University Research Foundation, August 10, 2017.
11. S. Fomine and M.A. Tlenkopatchev, Metathesis transformations of terpenes. computational modeling of (-)-α-pinene ring opening by ruthenium and tungsten carbene catalysts, *Journal of Organometallic Chemistry* 68–74 (2012).
12. M. Scharfenberg, J. Hilf, and H. Frey, Functional polycarbonates from carbon dioxide and tailored epoxide monomers: degradable materials and their application potential, *Advanced Functional Materials* 1704302 (2018).
13. F. Ng, G. Couture, C. Philippe, B. Boutevin, and S. Caillol, Bio-based aromatic epoxy monomers for thermoset materials, *Molecules* 149 (2017).
14. N.D. Sachinala and M.H. Litt, Epoxy monomers from sucrose, US Patent 5 646 226, assigned to Hawaii Agriculture Research Center, July 08, 1997.
15. A.V. Davis and P.N. Nickias, Zinc catalyst/additive system for the polymerization of epoxide monomers, US Patent 20 160 289 380, assigned to Dow Global Technologies LLC, October 06, 2016.

16. A.M. Tatara, E. Watson, T. Satish, D.W. Scott, D.P. Kontoyiannis, P.S. Engel, and A.G. Mikos, Synthesis and characterization of diol-based unsaturated polyesters: Poly (diol fumarate) and poly (diol fumarate-co-succinate), *Biomacromolecules* 1724–1735 (2017).
17. A.K. Mohanty, F. Wu, R. Mincheva, M. Hakkarainen, J.-M. Raquez, D.F. Mielewski, R. Narayan, A.N. Netravali, and M. Misra, Sustainable polymers, *Nature Reviews Methods Primers* 46 (2022).
18. J. Son, Y.J. Sohn, K.-A. Baritugo, S.Y. Jo, H.M. Song, and S.J. Park, Recent advances in microbial production of diamines, aminocarboxylic acids, and diacids as potential platform chemicals and bio-based polyamides monomers, *Biotechnology Advances* 108070 (2022).
19. K.S. Hayes, Polymerization of fatty acids, US Patent 4 776 983, assigned to Union Camp Corp, October 11, 1988.
20. G.R. Palmese, J.J.L. Scala, and J.M. Sands, Fatty acid monomers to reduce emissions and toughen polymers, US Patent 7 524 909, assigned to Drexel University, US Department of Army, April 28, 2009.
21. G.R. Palmese, J.J. LaScala, and J.M. Sands, Fatty acid monomers to reduce emissions and toughen polymers, US Patent 8 372 926, assigned to Drexel University, US Department of Army, February 12, 2013.
22. D. Swern, G.N. Billen, T.W. Findley, and J.T. Scanlan, Hydroxylation of monounsaturated fatty materials with hydrogen peroxide, *Journal of the American Chemical Society* 1786–1789 (1945).
23. F.D. Gunstone, *Fatty Acid and Lipid Chemistry*, Springer, (2012).
24. C. Litchfield, *Analysis of Triglycerides*, Elsevier, (2012).
25. E.N. Frankel, L.M. Smith, C.L. Hamblin, R.K. Creveling, and A.J. Clifford, Occurrence of cyclic fatty acid monomers in frying oils used for fast foods, *Journal of the American Oil Chemists' Society* 87–90 (1984).
26. G. Dobson, Cyclic fatty acids: qualitative and quantitative analysis in *Lipid analysis in Oils and Fats* 136–180. Springer, (1998).
27. W.W. Christie, Cyclopropane and cyclopropene fatty acids, *Topics in Lipid Chemistry* 1–49 (1970).
28. D.E. Minnikin, Location of double bonds and cyclopropane rings in fatty acids by mass spectrometry, *Chemistry and Physics of Lipids* 313–347 (1978).
29. T. Řezanka and V.M. Dembitsky, Identification of unusual cyclopropane monounsaturated fatty acids from the deep-water lake invertebrate Acanthogammarus grewingkii, *Comparative Biochemistry and Physiology Part B: Comparative Biochemistry* 407–413 (1994).
30. J. Sébédio and A. De Rasilly, Analysis of cyclic fatty acids in fish oil concentrates, in *17th Nordic Lipid Symposium* 212–216, (1993).

31. B. Hippchen, A. Röll, and K. Poralla, Occurrence in soil of thermo-acidophilic bacilli possessing ω-cyclohexane fatty acids and hopanoids, *Archives of Microbiology* 53–55 (1981).
32. Y.B. Tee, R.A. Talib, K. Abdan, N.L. Chin, R.K. Basha, and K.F.M. Yunos, Comparative study of chemical, mechanical, thermal, and barrier properties of poly (lactic acid) plasticized with epoxidized soybean oil and epoxidized palm oil, *BioResources* 1518–1540 (2016).
33. B.M. Malofsky and C. Mariotti, Methylidene malonate process, US Patent 20 150 183 714, assigned to OptMed Inc, July 02, 2015.
34. F.C. Tyler, Automatic feed-hopper for nut-tapping and other machines, US Patent 927 902, assigned to Individual, July 13, 1909.
35. A. Palsule, A.R. Holzer, P.R. Stevenson, K.K. Parab, and J.M. Sullivan, Solution polymers formed from methylene malonate monomers, polymerization, and solution polymer products, US Patent 20 170 240 657, assigned to Sirrus Inc, August 24, 2017.
36. M.N. Leiske and K. Kempe, A guideline for the synthesis of amino-acid-functionalized monomers and their polymerizations, *Macromolecular Rapid Communications* 2100615 (2022).
37. C. Bao, B. Kauffmann, Q. Gan, K. Srinivas, H. Jiang, and I. Huc, Converting sequences of aromatic amino acid monomers into functional three-dimensional structures: Second-generation helical capsules, *Angewandte Chemie International Edition* 4153–4156 (2008).
38. D.H. Martin, Rigid ring amino acids and polyamides there from, US Patent 7 098 361, assigned to Martin, Donald H, August 29, 2006.
39. M.A. Tapsak, Adhesives and methods for their manufacture and use, US Patent 8 937 140, assigned to Empire Technology Development LLC, January 20, 2015.
40. Wikipedia contributors, Isosorbide — Wikipedia, the free encyclopedia, https://en.wikipedia.org/wiki/Isosorbide, (2023).
41. C. Hoshino, T. Yamada, D. Taneda, Y. Nagata, T. Fujii, T. Mase, and Y. Ueno, Method for producing monosaccharides from biomass and monosaccharide production device, US Patent 20 070 148 750, assigned to JGC Corp., June 28, 2007.
42. Wikipedia contributors, Carbohydrate — Wikipedia, the free encyclopedia, https://en.wikipedia.org/wiki/Carbohydrate, (2023). [Online; accessed 13-August-2023].
43. Wikipedia contributors, Nucleotide — Wikipedia, the free encyclopedia, (2022). [Online; Stand 16. Juli 2023].

2

Automotive Industry, Hemp and Sustainable Polymers

2.1 Plastics Industry

The Fourth Industrial Revolution, also known as 4IR or Industry 4.0, conceptualizes the rapid change to technology, industries, and societal patterns and processes in the 21st century due to increasing interconnectivity and smart automation (1).

The term was popularized in 2015 by Klaus Schwab, Founder and Executive Chairman of the World Economic Forum, and has since been used in numerous economic, political, and scientific articles in reference to the current era of emerging high technology. Schwab asserts that the changes seen are more than just improvements to efficiency, but express a significant shift in industrial capitalism (1).

A study used bibliometric analysis to investigate the trends and the knowledge in Industry 4.0 research (2). Overall Industry 4.0 research was compared with that focused on the plastics industry by using data collected from current literature in the field. The analysis was conducted using the Bibliometrix R-tool and VOSviewer software. The results indicated that:

1. *Internet of Things* was the most commonly used keyword,
2. Industry 4.0 research can be divided into clusters,
3. Countries such as China and the USA dominate Industry 4.0 research,

4. Industry 4.0 research in the plastics industry is in its initial stages but is growing, and
5. There is a large difference between overall research and that in the plastics industry.

These results, including limitations and recommendations for future research, were discussed.

Industry 4.0 was introduced by the German Federal Government to refer to the current trend of establishing intelligent products and production processes and is commonly referred to as the Fourth Industrial Revolution, smart factory, intelligent factory, Internet of Things (IoT), and cyber physical system (CPS) (3).

The transition to Industry 4.0 is a priority for many practitioners and researchers (4, 5) and is reshaping industries by promoting the integration of industrial and systems information with communication technologies (6, 7).

Several factors indicate the importance of Industry 4.0, particularly its potential to address economic opportunities, social issues, and ecological issues (2).

Detailed research on bridging the gap between the theoretical and practical approach of Industry 4.0 has been presented (8). In particular, an established Greek company in the textiles and plastics industry was studied to present the current technological status of the industry. The following methods were checked (8):

1. Industrial process of a plastic manufacturing plant presenting the processes, methods, and equipment classification of such a factory.
2. Then, a new method for developing or upgrading factories specializing in textiles and plastics that showcase its operational procedures.

Also, the key points that need to be changed to develop the technological operations of an industry plan from Industry 3.0 to Industry 4.0 were identified. To achieve this, a detailed techno-economic proposal has been provided that suggests a trade-off between optimal quality and continuity of operations for an Industry 4.0 level factory.

Here, the technical novelty lies in taking a step back and providing a top-down technological approach to the current state of affairs in machinery, supply chain, and factory operations level while suggesting cost-effective methods to increase productivity, reduce production errors and pave the way for smart manufacturing techniques (8).

2.2 Fields of Application

Polymers are used for various fields of application. In Table 2.1 an overview is given.

Table 2.1 Applications of plastics.

Field	Special uses
Automotive	Cars
Food	Bottles, Packaging
Medical	Drug delivery, Artificial muscles, Tissue engineering
Electrical	Electrolytes, Electron transport, Cars, High temperature materials, Integrated cicuits, Solar cells, Wire isolation
Electronics	Photodiodes, Semiconducting polymer, Ferroelectric field effect transistors

2.3 Evolution

The issues of evolution, revolution and future of plastics for an eco-friendly environment have been described in a chapter of a book (9).

The evolution of plastics consists of the following steps (10):

1862 Birth of plastics
1880 Cellulose
1880 Celluloid
1907 Bakelite
1914 Poly(vinyl chloride)
1921 Injection molding
1922 Super polymers
1933 Polythene
1935 Aircraft cockpit

1950 Plastics in clothing industry
1956 Plastics in automobiles
1960 Plastics in furniture
1970 Plastics in telecommunications
1980 Plastics in nanotechnology
1990 Bioplastics
2021 Recycling of plastics

2.4 Material Safety

The positive benefits afforded by the widespread use of plastics need to be reconciled with the negative impacts on the environment and health across the entire plastics life cycle (11). Optimizing the balance in several facets of plastics production, use and waste management is necessary for a more sustainable relationship with these materials in the Anthropocene.

Plastics have brought us enormous societal benefits owing to their versatility, light weight, durability and low costs. However, these properties have come with negative externalities, especially because these persistent materials are leaked into the environment across their entire life cycle (11).

2.5 Environmental Sustainability of Plastics

The issues of plastics and environmental sustainability have been discussed in a book (12).

Nanocellulose has been described as a new material for environmental sustainability (13).

An overview of the current state of the science in nanocellulose research has been given.

The extraction procedures from lignocellulosic biomass have been detailed.

The process used to obtain nanocellulose from lignocellulosic biomass greatly influences its morphology, properties and surface chemistry.

The efficacy of chemical methods that use alkali, acid, bleaching agents, ionic liquids, and deep eutectic solvent for pretreatment of biomass have been discussed.

There has been a continuous endeavour to optimize the pretreatment protocol as it is specific to lignocellulosic biomass and also depends on factors such as the nature of the biomass, process and environmental parameters and economic viability.

Nanofibers are primarily isolated through a mechanical fibrillation process. Nanocrystals are predominantly extracted using an acid hydrolysis process.

Also, techniques have been described that can be used to modify the surface properties of nanocellulose by functionalizing surface hydroxyl groups to impart desirable hydrophilic-hydrophobic balance.

The demand for cellulose is expected to increase further due to its increasing and broadening applications (13).

2.6 Future Directions for Sustainable Polymers

2.6.1 Hemp

Most plastics produced today are made using petroleum-based compounds that release harmful gases into the atmosphere. Waste solutions are inefficient, and harmful by-products are toxic to our land, water and wildlife (14).

Hemp plastic is a plastic that's 100% biodegradable in nature if produced by using only hemp plant. A hemp plant consumes four times more carbon dioxide from atmosphere than other plants. The fibers produced from hemp are the conventional fibers currently being used. Numerous benefits of using hemp for the manufacturing of biodegradable plastic rather than conventional plastics have been detailed (14).

The strength of hemp plastic makes it appealing in several industries. The automotive, building, and packaging industries are all being turned on to hemp plastic. It is also lightweight and has a very high ratio of density to weight. This allows it to potentially be used in aerospace to save weight on heavy structures. Due to its versatility, hemp can be used anywhere for any purpose, replacing conventional plastics.

Although plastic materials produced using petroleum compounds are quite cheap and easy to manufacture, the processes involved in the production of biodegradable plastics, or specifically hemp plastics, are quite expensive require more effort to produce than the production of conventional plastics (14).

Global warming has recently drawn the attention of the entire conscious community, with the fear that if not addressed properly, it will result in the extinction of numerous species around the world (15). At the same time, it will pose a threat to human health, food security, the living environment and standard of living. Therefore, possible solutions are being explored accordingly; regulations are being imposed in places to bind green production practices and limit the emission of CO_2, and an emphasis is being placed on renewable resources along with the search for alternatives to carbon-positive materials.

Hemp has received a lot of attention because of its multipurpose usability, short production cycle, low capital demand in cultivation, possibility of carbon-negative transformation and easy carbon sequestering material. A review of hemp as a very promising renewable resource, including its potential uses in paper, textiles, composites, biofuel, and the food industry, is presented in a paper (15).

Hemp is one of the ancient plants cultivated by humans for textile use. It is believed that hemp cultivation started in western Asia and gradually spread worldwide. Currently, more than 30 countries are involved in the global hemp trade (16).

Due to its capability of growing in pesticide and herbicide-free environments, a noticeable resistance to rodents, fungus and many types of weeds ocurs (17) for a wide geographical range of cultivation and multipurpose uses.

Three major hemp-producing countries by production area are Canada (555,853 *ha*), North Korea (21,247 *ha*), and France (12,900 *ha*) (18).

Hemp requires a similar soil preparation than other break crops with a pH of 6.0 to 7.5 (19). It grows well on loose, well-drained loam soils with a mean temperature between 16°C to 27°C and high moisture. The final yield depends on sowing density, nitrogen level, harvest time, and it can grow up to 0.31 *m* in a week (20).

Hemp is a tall, annual crop with a low labor-intensive production process (21), which can be accomplished in a short cropping period (70–90 *d*). Hemp cultivation has become more appealing to farmers than flax because it has a lower chance of crop failure. It guarantees higher yields (up to 12 tons per hectare as cellulose, 20 tons as stem particles and 25 tons as fiber matter per hectar) while also enhancing soil nutrition (22, 23).

The industrial norms for producing this crop are generally guided by organic cultivation to maintain the fiber quality in terms of fineness, strength and color. Hemp fiber cultivation requires about 77.63% less cost in fertilization, seeds, field operation, and irrigation costs than cotton, the most recognized natural fiber (24). It shows many unique properties that differ from other natural fibers due to their aseptic properties, high absorbency, protection against UV radiation, and allergen-free effect (25).

2.6.1.1 *Hemp Biocomposites*

The combined effect of design parameters on the mechanical properties of the biocarbon-filled hemp polymer composite offers a newer dimension in biocomposite research. Biopolymer composites with or without hemp fiber and biocarbon fillers were fabricated and optimized (26). The results showed biocarbon with a particle size of 50 μm and a filler loading of 10%. A pyrolysis temperature of 650°C showed the maximum tensile strength (840.75 MPa with switchgrass biocarbon; 817.02 MPa with hemp biocarbon).

The tensile strength of the composite samples was directly proportional to the particle filler loading. The tensile strength initially improved by 50% when particle size increased to 75 μm. A further increase reduced the strength of the composites.

The energy at tensile rupture increased with particle size. In contrast, the increased filler loading was detrimental to the energy at break. The parameters positively impacted the flexural strength of the samples. The impact strength of the samples fell by 63% when filler loading was doubled to 20% (26).

Research was conducted on the hygroscopicity, mechanical properties, and biofouling resistance of biocomposites made of epoxy resin with 28 m% bio-based carbon content reinforced with flax and hemp fibers (27). A series of experiments were performed to acquire water absorption rates, saturation limits, mass increase, tensile and flexural properties, interlaminar shear strength, impact resistance, and mass gain due to biofouling. All tests of mechanical properties were conducted before and after immersion in seawater. The acquired saturation limits of flax/epoxy and hemp/epoxy biocomposites amounted to 7.5% and 9.8%, respectively. The water uptake caused the tensile and flexural properties to decrease by 26–74%, while interlaminar and impact strength increased for flax/epoxy and decreased for hemp/epoxy biocomposites (27).

In addition, it was observed that in almost all cases, flax/epoxy had superior properties compared with hemp/epoxy biocomposites. It is expected that this research will motivate naval architects and classification societies to consider biocomposites as prospective hull materials that provide both structural integrity and environmental attainability (27).

2.7 Circular Economy

Circular economy is a model of production and consumption which involves sharing, leasing, reusing, repairing, refurbishing and recycling existing materials and products for as long as possible (28).

Circular economy aims to tackle global challenges such as climate change, biodiversity loss, waste, and pollution by emphasizing the design-based implementation of the three base principles of the model. The three principles required for the transformation to a circular economy are:

1. Designing out waste and pollution,
2. Keeping products and materials in use, and
3. Regenerating natural systems.

In a circular economy, business models play a crucial role in enabling the shift from linear to circular processes. Various business models have been identified that support circularity, including product-as-a-service, sharing platforms, and product life extension models. These models aim to optimize resource utilization, reduce waste, and create value for businesses and customers alike, while contributing to the overall goals of the circular economy.

Circular economy strategies can be applied at various scales, from individual products and services to entire industries and cities (28).

2.7.1 Food Waste

In a study, the key gaps of food waste prevention have been addressed in the context of the emerging circular economy (29). First, current terminology related to food waste was reviewed and clarified, in particular, the terms food surplus, waste and losses. This work highlights why the clarity of these definitions is crucial for the sustainability of future food waste management systems, especially in the context of circular economy. Through a simple matrix, definitions are linked to the concepts of edibility and possibility of avoidance, leading to six distinct categories of food waste (29):

1. Edible,
2. Naturally inedible (pits),
3. Industrial residue,
4. Inedible due to natural causes (pests),
5. Inedible due to ineffective management, and
6. Not accounted for.

An updated pyramid for food waste hierarchy was proposed, distinguishing surplus food and a new category for material recycling, in order to reflect the future food waste biorefineries in the circular bioeconomy. Nutrient and energy recovery are two separate categories and the terms recovery and recycling are clarified. Finally, a circular economy framework was presented for food surplus and waste, considering closing the loop throughout the whole food supply chain, in connection with the concept of strong and weak sustainability (29).

About 30% of the food produced globally is lost or wasted along the food supply chain. So, the objective of a study was to identify and systematize the reasons relating to food loss and waste with solutions based on the concept of circular economy (30).

Based on a single case study analysis, a study quantified the food loss and waste along the supply chain of Italian pasta, emphasizing food loss and waste valorization from a circular economy perspective using the global Food Loss and Waste Accounting and Reporting Standard (31).

The Food Loss and Waste Accounting and Reporting Standard was developed in 2016 (32).

The results of the study showed that the pasta supply chain is a good example of circular economy as little is lost. Food losses in the field are less than 2%, while the straw obtained during the harvest is normally used as animal feed and litter. The losses generated during the grinding of the wheat and pasta production amounted to approximately 2%. In line with the previous literature, most food loss and waste occurs during the cultivation and consumption, thus demonstrating that further research is required in order to reduce food loss and waste in these two supply chain phases (31).

Around a third of the food produced globally is wasted as a result of the significant resources required for its treatment and disposal, which also wastes valuable resources (33). Following the circular economy principles, this waste should ideally be avoided, and if not possible, treated to recover resources. Four treatment methods were analyzed (33):

1. Anaerobic digestion,
2. In-vessel composting,
3. Incineration, and
4. Landfilling.

The results of this study showed that per tonne of waste treated, anaerobic digestion has the lowest environmental impacts in 13 out of the 19 categories considered in the study, including net-negative global warming potential. In-vessel composting is the least sustainable option environmentally, in contrast to being preferred over incineration according to the circular economy principles. Incineration has the lowest life cycle costs (£71 per tonne), while landfilling is the costliest option (£123 per tonne). Managing the 4.9 *Mt* of food waste collected annually from UK households via the four methods generates 340,000 *t* CO_2.

If all of this food waste was incinerated, a lot could be saved, compared to the common waste management, rendering incineration a carbon-negative technology. This would also result in savings in 14 other impacts, but would increase summer smog by 30% and metal depletion by 56%. The environmental benefits of incineration would be exceeded only if all food waste was treated by anaerobic digestion.

Anaerobic digestion would also lead to savings in 14 other impacts compared to the present situation, but would result in a four times higher acidification and three times greater emissions of particulate matter (33).

Although plastic is one of the most commonly used materials in everyday life, the current linear economy (produce, use and dispose) engenders high risks to human health in relation to greenhouse gas emissions and environmental pollution (34). As a response to these challenges, the circular plastic economy is gaining momentum, where the goal is to reduce, reuse and recycle all plastic.

The transition to the circular economy should be made across the entire plastics value chain in order to ensure circular design, production, use and waste management. This study examines the current scientific literature in relation to the entire value chain of plastics (34). An overview has been given of the existing research associated with the transition of plastic use to a circular model. The literature was divided into the following categories (34):

1. Design,
2. Production,
3. Use,
4. End-of-life, and
5. Value chain.

A high proportion of the literature was found to address the end-of-life phase, suggesting that the other phases are currently neglected. The results have implications that are applicable to multiple phases.

In particular, contamination of waste streams and composite materials places significant limitations on the opportunity to recycle and reuse plastic in new products. This calls for changes in the whole value chain, and for trans-sectorial collaboration to ensure systemic transparency. Therefore, future research should take a holistic approach to the transition to a circular economy through careful mapping of implications, stakeholder involvement and collaboration (34).

A circular economy approach to plastic waste was presented in a study (35). Growing environmental concerns associated with the accumulation of plastic waste in the natural environment has led to research of renewable alternatives, and more recently, alternative waste management strategies.

Recent life-cycle assessments suggest that poly(lactic acid) (PLA) offers reductions of up to 40% in greenhouse gas emissions and 25% in nonrenewable energy use compared to traditional petrochemical-based plastics, including poly(ethylene) (PE) and poly(ethylene terephthalate), thus exemplifying the green credentials of PLA. Chemical recycling of PLA has attracted increasing interest in recent years owing to its potential to further integrate PLA into a circular economy.

Presently, the chemical recycling options available for PLA can be characterized into two distinct routes, namely depolymerization or degradation. Although the terms depolymerization and degradation are often used interchangeably, it is important to recognize there is a distinct technical difference (35).

2.7.2 *Conversion of Waste Plastic into a Feedstock*

A continuous process was described for converting waste plastic into a feedstock for PE polymerization. The process consists of the selection of waste plastics containing PE and/or poly(propylene) (PP), and then passing the waste plastics through a pyrolysis reactor to thermally crack at least a portion of the polyolefin waste and produce a pyrolyzed effluent.

The pyrolyzed effluent is then separated into offgas, a pyrolysis oil comprising a naphtha/diesel/heavy fraction, and char. The pyrolysis oil is passed to a crude unit in a refinery from which a naphtha fraction (C_5–C_8), or a propane and butane fraction, is recovered. The naphtha fraction, or propane and butane fraction, is then passed to a steam cracker for the ethylene production.

Figure 2.1 shows a diagram of the pyrolysis of plastic waste into fuel or wax that is generally used in the industry. PE and PP wastes are sorted together 1. The cleaned PE/PP waste 2 is converted in a pyrolysis unit 3 to offgas 4 and pyrolysis oil (liquid product), and at times wax. The offgas 4 from the pyrolysis unit is used as fuel to operate the pyrolysis unit 3. An on-site distillation unit separates the pyrolysis oil to produce naphtha and diesel products 5 which are sold to fuel markets. The heavy pyrolysis oil fraction 6 is recycled back to the pyrolysis unit 3 to maximize the fuel yield. Char 7 is removed from the pyrolysis unit 3. The heavy fraction 6 is rich in long chain, linear hydrocarbons, and is very waxy (i.e., forms

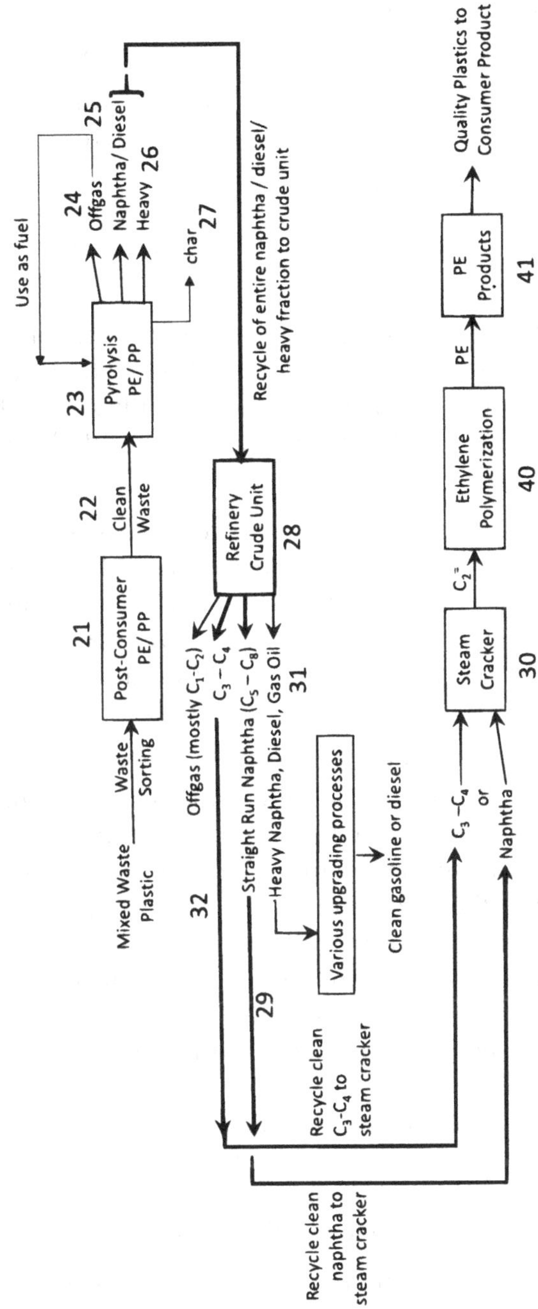

Figure 2.1 Pyrolysis of waste plastics (36).

paraffinic wax upon cooling to ambient temperature). The wax can be separated from the heavy fraction 6 and sold to wax markets (36).

This process converts pyrolyzed PE and/or PP waste plastic in large quantities by integrating the waste polymer pyrolysis product streams into an oil refinery operation (36, 37). The resulting processes produce the feedstocks for the polymers (naphtha or C3-C4 for ethylene cracker), high quality gasoline and diesel fuel, and/or quality base oil.

Generally, the present process provides a circular economy for PE plants. PE is produced via polymerization of pure ethylene. Clean ethylene can be made using a steam cracker. Either naphtha or a C_3–C_4 stream can be fed to the steam cracker. The ethylene is then polymerized to create PE.

By adding refinery operations to upgrade the waste pyrolysis oil and wax to higher value products (gasoline and diesel) and to produce clean LPG and naphtha for steam cracker for ultimate PE polymer production, one is able to create positive economics for the overall process from recycled plastics to PE product with quality identical to that of the virgin polymer.

A pyrolysis unit produces poor quality products containing contaminants, such as calcium, magnesium, chlorides, nitrogen, sulfur, dienes, and heavy components, which cannot be used in large quantity for blending in transportation fuels. It has been discovered that by having these products go through the refinery units, the contaminants can be captured in pre-treating units and their negative impacts diminished. The fuel components can be further upgraded with appropriate refinery units with chemical conversion processes, with the final transportation fuels produced by the integrated process being of higher quality and meeting the fuels quality requirements. The integrated process will generate a much cleaner naphtha stream for stream cracker feedstock for ethylene generation and for PE production. These large on-spec productions make a cyclical economy for the recycle plastics feasible.

The carbon in and out of the refinery operations is transparent, meaning that all the molecules from the waste plastic do not necessarily end up in the exact olefin product cycled back to the polyolefin plants (36).

2.8 Automotive Industry

The automotive industry is critical to economics, research and development, and global innovation growth (38). The global automotive industry fabricates more than 66 million cars (including vans, trucks, and buses) per year. According to the International Energy Agency (IEA) 2019 report, passenger cars (including cars, sport utility vehicles, and personal trucks) alone use more energy for their operation than the whole residential sector and use comparatively the same energy as the entire manufacturing industry.

The transportation sector accounts for 20.5% of the global energy consumption (39). The operational energy consumption in vehicles is directly related to weight, and weight is directly related to the materials. Therefore, it is important to use lighter materials to reduce energy consumption, but it is also essential that these materials are friendly to the environment.

The transport sector is responsible for almost a third of the final energy-related CO_2 emissions, of which 72% comes from road transportation, mostly from passenger cars (40).

Plastics typically make up 18% of the average weight of a new vehicle, with more than 50% of the volume (41).

Four areas require high-priority lightweighting research and development with plastics: interior, body, powertrain, and chassis. Additionally, bioplastics and composites are good candidates to substitute metals and metal alloys to manufacture several automotive components.

The importance of plastics is recognized in reducing weight, energy consumption, or CO_2 emissions, but above all, there is a need for sustainable plastics in the automotive industry.

In the transportation sector, the automotive industry has used plastics almost since the beginning. The lightness, flexibility, and many qualities of plastics make them ideal for the automotive industry, reducing the overall weight of cars. This results in less fuel consumption (42).

Typically, plastics are used in exterior parts such as body panels, seals, wheel covers, weatherstripping, bumpers, fenders, air dams, and trims; and interior features such as the instrument panel, dashboard, door panels, steering wheels, seat and associated parts, instrument panel skin, and decorative pieces. Under-hood components such as sensors, ignition compartments, fluid systems, power distribution, and resonators also use plastics (43).

Engineering plastics, such as polyamide (PA), polyphenylene sulfide (PPS), poly(methyl methacrylate) (PMMA), and acrylonitrile butadiene styrene (ABS), among others, belong to the high-performance segment of nonrenewable resources. These plastics exhibit higher properties than commodity plastics such as poly(ethylene) (PE), poly(propylene) (PP), poly(vinyl chloride) (PVC), and poly(styrene) (PS). They exhibit excellent strength, good temperature resistance, toughness, stiffness, chemical resistance, light weight, wear and abrasion resistance, and they easily make automobile components (44).

In some cases, plastics can easily substitute metal components. The use of engineering plastics goes from single plastics to more sophisticated copolymers, such as the sunroof systems of Renault (a 15% glass-reinforced copolymer compound of styrene maleic anhydride and acrylonitrile butadiene styrene (SMA-GF)) and Citroen (Xiran glass-reinforced blend of styrene maleic anhydride (SMA) and ABS) (45). The plastics most commonly used in the automotive industry include acrylonitrile butadiene styrene (ABS), polyamide (PA), polycarbonate (PC), poly(ethylene terephthalate) (PET), poly(styrene) (PS), and poly(butylene terephthalate) (PBT). Table 2.2 summarizes the different nonrenewable resource-derived plastics currently used; some have been used in low-volume production (46–48)

Common plastics used in a typical car are shown in Table 2.2.

Fossil-derived plastics are not biodegradable by microorganisms within a reasonable time frame. Generally, it would take about 300 years for 60 *mm* of some plastic films to degrade entirely in soil; this is why plastics are considered an ecological problem (49). Although capable microorganisms and engineered enzymes make the microbial degradation of polymers such as PET possible (50, 51), fossil-derived plastics are not compostable.

The usage of renewable-sources-derived plastics is a step toward the current environmental goals set for the automotive industry. The substitution of some fossil-fuel-derived plastics adds to the solution of lightweighting, a much-needed solution to waste management, and solves industrial and ecological issues surrounding plastic dis-

Table 2.2 Common plastics used in a typical car.

Component	Polymer Types
Bumpers	PS, ABS, PC/PBT, PP, PA, PU, TPO
Fascia systems	PS, ABS, PC/PBT, PP, PA, PU, TPO
Seating	ABS, PA, PP
Instrument panels	ABS, PC, ABS/PC, PP
Fuel systems	POM, PA, PBT
Under hood components	PA, PBT
Interior trim	ABS, PET, POM
Electrical components	PBT, PA
Exterior trim	PS, PVC, ABS, PA, PBT, POM, ASA
Lighting systems	PC, PBT, ABS, PMMA
Upholstery	ABS, PU
Liquid reservoirs, cooling	PA
Battery carriers	PA
Wheel covers	ABS
Body parts	ABS
Tires	PA
Parts of engine	PA, phenolic resins

Abbreviation	Full name
ABS	Acryl butadiene styrene
ASA	Acrylonitrile styrene acrylate
PA	Poly(amide)
PBT	Poly(butylene terephthalate)
PC	Poly(carbonate)
PET	Poly(ethylene terephthalate)
PMMA	Poly(methyl methacrylate)
POM	Poly(oxymethylene)
PP	Poly(propylene)
PS	Poly(styrene)
PU	Poly(urethane)
TPO	Thermoplastic poly(olefin)s

posal. Bioplastics, natural-fiber composites, and fiber-reinforced polymer composites are the current alternative. Because there is no consensus on the definitions, descriptions are provided along with the extent of their use in automotive applications.

Bioplastics are a family of bio-based, biodegradable materials, or both. Bio-based plastics are human-made or processed organic macromolecules from renewable sources, such as corn, potatoes, wheat, and vegetable oil, through chemical or biological processes (52).

Bio-based plastics, often called biopolymers, can be produced by direct extraction from natural substances, the polymerization of monomers derived from biomass, or by microorganisms (53).

Natural biopolymers produced intra- or extra-cellularly in living organisms include cellulose, chitin, starch, polyhydroxyalkanoates (PHA) such as polyhydroxybutyrate (PHB), and synthetic biopolymers built via synthesis reactions, such as poly(lactic acid) (PLA) and others (54).

Bioplastics are biodegradable or compostable. A biodegradable polymer undergoes biodegradation, a chemical process during which microorganisms in the environment decompose materials into natural substances such as water, carbon dioxide, and methane (52).

Cellulosic plant fibers are directly isolated from plants such as jute, kenaf, hemp, flax, sisal, banana, bamboo, or coir (coconut fiber) (55).

Renewable plant fibers can replace non-degradable fibers in fiber-reinforced plastic composites for lightweighting (56). Typical fiber-reinforced plastic composites (such as glass- and carbon-fiber-reinforced plastics) combine the high mechanical and physical performance of the fibers and the appearance and physical properties of the polymers (such as polyester or PP) (55, 57).

The lack of biodegradability of fiber-reinforced composites constitutes a significant disadvantage. Thus, substituting industrial fibers with natural fibers is desirable because they have low cost, low environmental impact, and relatively comparable properties to some metals and other composites (58).

Commodity plastics, such as poly(propylene) and poly(ethylene), and engineering plastics, such as polycarbonate, polyamides, and poly(styrene), are the primary polymer matrix in these composites (59, 60).

Another advantage of using natural fibers for composite production is reducing CO_2 emission (61, 62).

When decomposed, the amounts of CO_2 that the natural fiber releases are the same as what the plant assimilates during the growth phase (63, 64).

Highly biodegradable polymers (also known as bioengineering polymers, biopolymers, or bioplastics), such as certain aliphatic polyesters (i.e., PHA), starch-based, and protein-based, are highly biodegradable and carbon neutral (65).

Highly biodegradable polymers help design eco-friendly products to satisfy customer demand and increasingly strict legislation (66).

However, they are still incapable of many long-service-life applications. Most biodegradable product applications focus on short-service-life products (i.e., bags, disposable plates and cutlery, packaging materials, and food containers). For automotive applications, newer production technologies to enable the economy of scale are necessary for mass production and must address the challenges of preventing biopolymer degradation during the lifetime of the car. The National Highway Traffic Safety Administration investigated opportunities for lightweight vehicles using advanced plastics and composites in 2012 (48). They found that using structural plastics and composites will make lighter, more fuel efficient, and environmentally sustainable vehicles. After modeling and simulation, they concluded that the use of 30% in the content of plastics and composites would have a weight reduction effect without adverse effects on the vehicle crashworthiness. However, bioplastics are still limited in automotive applications primarily because of their low mechanical properties. Additionally, some polymers undergo thermal degradation at low temperatures (67), which means that on a sunny day (where the temperature inside the car reaches 50°C to 70°C), the polymers could emit an undesirable odor because of their early degradation. Those technical and processable drawbacks hamper their wide-range applicability (68). Nevertheless, reinforcing helps overcome these drawbacks by blending or adding other polymers or fillers. Thus, biopolymers can significantly improve the final mechanical properties, thermal behavior, and degradation mechanisms through a cost-effective, accessible, and readily available processing technology (69, 70).

The most used bioplastics in the automotive industry include naturally occurring fibers such as soy and hemp, bio-polyamides (bio-PA) and their composites, DuPont Zytel, a combination of nylon resin materials, poly(lactic acid) (PLA), and bio-based poly(propylene) (bio-PP) (71).

The properties and characteristics of the composite biodegradable material satisfy the increasingly demanding applications that the biodegradable polymer alone could not. All major vehicle manufacturers worldwide use bioplastics and bio-based plastics, including natural-fiber composites and engineering plastics reinforced with natural fibers such as flax, hemp, jute, and sisal

Natural-fiber-reinforced polymer composites used in automotive applications are shown in Table 2.3.

Additionally, Ford vehicles now boast eight sustainable-based materials, including soy foam, wheat straw, kenaf fiber, cellulose, wood, coconut fiber, rice hull, and agave fibers from tequila industry waste.

Ford is aiming to get rid of single-use plastics by 2030.

Moreover, Toyota sustainable development goals by 2050 include the challenge of establishing a recycling-based society and systems whose primary purpose is to reduce the consumption of dwindling natural resources through the use of renewable resources and recycled materials. The goals are to reduce petroleum-based plastics by developing recycled and eco-plastic technology, meeting quality and performance requirements and establishing collection systems for used plastics (72).

Audi aims to recycle mixed automotive plastic waste in a resource-conserving closed loop (38).

2.8.1 Airbag

An airbag apparatus for the front seat or driver seat of an automobile or mounted on the side pillar of the automobile basically includes an airbag, an airbag case for accommodating the folded airbag, and an inflater for inflating the airbag. The airbag apparatus is disposed inside an interior surface panel of the automobile. A typical example of such an interior surface panel is an instrument panel cover which

Table 2.3 Natural-fiber-reinforced polymer composites.

Company	Parts
Audi	Seat back, side, and back door panel, boot lining, hat rack, spare-tire lining
Citroen	Interior door paneling
BMW	Door panels, headliner panel, boot lining, seat back, noise insulation panels, molded footwell lining
Lotus	Body panels, spoiler, seats, interior carpets
Fiat	Door panel
Opel	Instrumental panel, headliner panel, door panels, pillar cover panel
Peugeot	Front and rear door panels
Rover	Insulation, rear storage shelf/panel
Toyota	Door panels, seat backs, floor mats, spare tire cover
Volkswagen	Door panel, seat back, boot-lid finish panel, boot liner
Mitsubishi	Cargo area floor, door panels, instrumental panels
Mercedes-Benz	Door panels, windshield/dashboard, business table, pillar cover panel, glove box, instrumental panel support, insultation, molding rod/apertures, seat backrest panel, trunk panel, seat surface/backrest, internal engine cover, engine insulation, sun visor, bumper, wheel box, roof cover
Honda	Cargo area
Volvo	Seat padding, natural foams, cargo floor tray
General Motors	Seat backs, cargo area floor
Saturn	Package trays and door panel
Ford	Floor trays, door panels, B-piller, boot liner

is made of a thermoplastic resin material, such as poly(propylene), and which covers the surface of an instrument panel core formed integrally from a thermoplastic resin material such as poly(propylene) (PP). Conventionally, a seamless-type airbag apparatus for a front passenger seat has been provided.

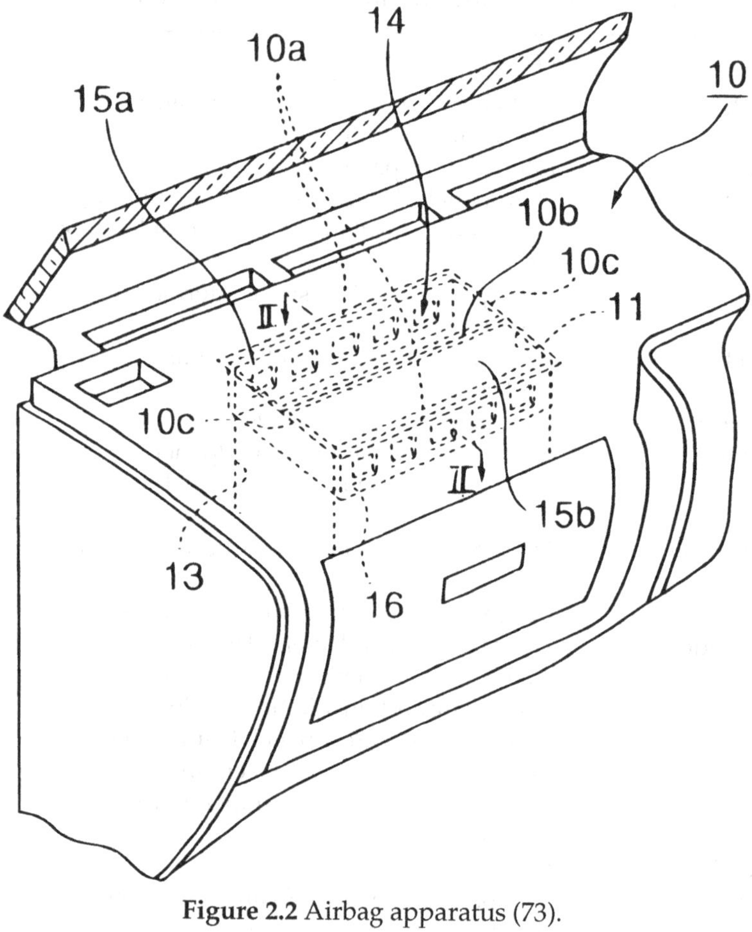

Figure 2.2 Airbag apparatus (73).

As shown in Figure 2.2, the seamless-type airbag apparatus is designed to render a fracture groove of a fracture-opening section of an instrument panel cover invisible. Specifically, the instrument panel cover 10 has a fracture-opening section 14 having a size corresponding to that of an opening portion 13 of an airbag case. The fracture-opening section 14 is formed through provision of front and back hinge grooves 10a extending longitudinally, a center fracture groove 10b extending longitudinally, and side fracture grooves 10c extending transversely. These grooves 10a, 10b, and 10c are formed on the inside surface of the instrument panel cover 10 by use of a laser. Specifically, a pulsating laser beam is perpendicularly applied to the inside surface of the instrument panel cover 10 while being moved along peripheral edges of a pair of reinforcement plate members 11 fixed to the inside surface, thereby forming the grooves 10a, 10b, and 10c. The longitudinally extending center fracture groove 10b and the side fracture grooves 10c formed perpendicularly to the fracture groove 10b divide the fracture-opening section 14 into a front fracture-opening subsection 15a and a rear fracture-opening subsection 15b. Upon inflation of an airbag, the fracture-opening subsections 15a and 15b are split apart from each other in a casement condition (in opposite directions) while hinging on the front and rear hinge grooves 10a and 10b. When, upon inflation of an airbag, the fracture-opening section 14 of the instrument panel cover 10 is split open by pressure of the inflating airbag 12, the front and rear hinge groove 10a may be broken, with a resultant separation of the fracture-opening subsections 15a and 15b from the instrument panel cover 10. In order to avoid this problem, the conventional air bag apparatus employs a reinforcing structure including the above-mentioned paired reinforcement plate members 11, which are provided on the inside surface of the fracture-opening section 14 of the instrument panel cover 10. Specifically, a frame-shaped reinforcement rib 16 is formed integrally with the instrument panel cover 10 in such a manner that the rib 16 projects from the inside surface of the instrument panel cover 10, and the rib 16 defines a rectangular space slightly larger than the opening portion 13 of the airbag case 13. The paired reinforcement plate members 11 are disposed within the rectangular space defined by the rib 16. The horizontal portions 11 of the reinforcement plate members 11 are fixedly attached to the inside surface of the fracture-opening section 14 through thermal joining or like processing. Vertical portions 11 of the reinforcement plate members 11 are bent at their hinge portions 11 and extend downward along the inner surfaces of front and rear walls of the reinforcement rib 16. The front and rear walls of the reinforcement rib 16 and the corresponding vertical portions 11 of the reinforcement plate members 11 are engaged with corresponding hooks 17 attached to front and rear walls of the airbag case 13. The reinforcement plates 11 prevent scattering of the fracture-opening subsections 15a and 15b of the fracture-opening section 14 of the instrument panel cover 10 upon inflation of the airbag (73).

The so configured airbag apparatus functions in the following manner. Upon collision of the automobile, an impact force caused by the collision is detected by a sensor. A control unit including a CPU judges whether or not the detected impact force is equal to or greater than a predetermined value. When the control unit judges that the impact force is equal to or greater than the predetermined value, the control unit issues a signal for causing the inflater to generate a predetermined gas. The gas is fed to the airbag so as to promptly inflate the airbag. The inflating airbag presses, from inside, the fracture-opening section 14 of the interior panel 10. The fracture-opening section 14 is fractured along the center fracture groove 10b and the side fracture grooves 10c, whereby the fracture-opening subsections 15a and 15b are split open about the front and rear hinge grooves 10a. The thus-opened front and rear fracture-opening subsections 15a and 15b are further opened while being turned inside out about the hinge portions 11 of the reinforcement plate members 11. Simultaneously, the inflating airbag 12 projects outward from the interior panel 10 through the thus-opened fracture-opening section 14. Serving as a cushion, the inflated airbag 12 supports a front seat passenger at his/her chest and head, thereby protecting the passenger from the impact force of collision.

2.8.2 *Biomass Pellets*

General plastic materials have stable molecular structures, so it is difficult to break the bonds and, even when they are broken, bonding tends to be formed again. For this reason, about 300 to 400 years are required for degradation of the plastic materials. As a result, soil pollution, scarcity of landfills, etc., are becoming severe problems (74).

In addition, air pollution resulting from the toxic gases and carbon dioxide produced during the burning of the plastics accelerates global warming.

Despite these problems, plastics are still used for automotive interior parts. Considering that about 57 million cars were produced in

2010, a considerably large amount of plastics are used as automotive interior parts.

Eco-friendly plastics, such as biodegradable plastics, have been described. Since plastics manufactured from modified starch exhibit poor physical properties, such as tensile and elongation properties, thermoplastic starch prepared by plasticization under high temperature and high pressure or PLA processed through starch fermentation and polymerization is used in practical applications. However, they do not meet the physical property requirements for automotive interior parts either. At present, only sample preparation is possible and large-scale production is difficult.

A plastic material for automotive interior parts has been described that uses a biomass pellet, which meets the physical property requirements for automotive interior parts. The biomass pellet can be prepared by injection molding a pellet containing a paper powder and a plastic resin. Furthermore, the pellet may contain a plant powder. Preferably, the pellet may contain 5-50% of the paper powder and 50-95% of the plastic resin based on total weight The plastic resin may contain PP, acrylonitrile-butadiene-styrene (ABS), poly(carbonate) (PC)-ABS, polyamide 6 and polyamide 66. The pellet may further contain an organic acid and peroxide, and a graft binder.

A plastic for automotive interior parts formed from a biomass pellet according to the present disclosure meets the physical property requirements for a plastic for automotive interior parts and is durable for 20-30 years.

The present disclosure is also advantageous in that recycling is easier using the biomass pellet, carbon dioxide emission can be reduced and the plastic is highly biodegradable after the disposal of waste (74).

2.8.2.1 *Method for Manufacturing*

Hereinafter, a method for manufacturing the plastic article using a plastic composition for automotive interior parts using a biomass pellet is described in detail (74).

In step 1, a paper powder is prepared by pulverizing and drying paper. Preferably, the paper may be pulverized into ultrafine powder of about 100-400 mesh. If the mesh size is smaller than

100 mesh, the particle size is too large, resulting in decreased pellet productivity due to poor flow property, unsatisfactory quality of the final product, such as film due to rough surface, and unsatisfactory strength and elongation of the product. Additionally, if the mesh size exceeds 400 mesh, it takes too long a time for the pulverization, resulting in decreased overall productivity and price competitiveness issues following increased production cost.

Then, in step 2, a plant powder is prepared by pulverizing and drying a plant. Preferably, the plant may also be pulverized into ultrafine powder of about 100-300 mesh. As with the paper powder, if the mesh size is smaller than 100 mesh, the particle size is too large, resulting in decreased pellet productivity due to poor flow property, unsatisfactory quality of the final product such as film due to rough surface, and unsatisfactory strength and elongation of the product. Additionally, if the mesh size exceeds 300 mesh, it takes too long a time for the pulverization, resulting in decreased overall productivity and price competitiveness issues following increased production cost.

In step 3, a biomass powder is formed by mixing the paper powder prepared in step 1 with the plant powder prepared in step 2.

In step 4, the biomass powder formed in step 3 is coated with a wax. A wax is added to the biomass powder which was formed through step 1 and step 3 and then dried, and a wax-coated biomass powder is formed by stirring at high speed. The wax melts by the heat generated during the stirring at high speed and is coated on the surface of the plant powder. The wax-coated biomass powder is prevented from reabsorbing moisture.

In step 5, a mixture is formed by mixing the wax-coated biomass powder formed in step 4 with a plastic resin which serves as a binder, an organic acid and peroxide. The peroxide serves to chemically break the chains of the plastic resin. The organic acid serves as an intermediate to bind the biomass powder to the terminal groups of the plastic resin cut by the peroxide and also as an agent of oxidative dissolution. In step 5, a graft binder may be further added in addition to the biomass powder, the plastic resin, the organic acid and the peroxide. The graft binder serves to help graft binding between the plastic resin and the biomass powder.

In step 6, a plastic resin-biomass graft binding composition is formed from the mixture formed in step 5 through graft binding between the plastic resin and the biomass by reacting in an extruder at 100-300°C. If the reaction temperature is below 100°C, reaction cannot occur because the added materials are not melted. Additionally, if the reaction temperature is above 300°C, carbonization may occur or preparation of a pellet may be impossible as the plastic resin is melted and flows like water.

In step 7, a pellet is prepared by discharging the plastic resin-biomass graft binding composition formed in step 6 through a discharge port, blow drying the discharged strand while transferring using a conveyor belt and then cutting by, for example, die-face hot cutting.

The biomass pellet that went through the above-mentioned steps may be injection molded to prepare a plastic article for automotive interior parts (74).

2.8.3 *Plastic Panels*

Traditionally, glazing has used glass material for automotive applications (75). Poly(carbonate) (PC) is a material that lends itself to this due to its characteristics for vehicle glazing. It is a clear-sighted material with extremely high impact strength, restorative properties after mechanical impact, high temperature resistance and high transparency. Also PC is very suitable for glazing in automotive vehicles, the properties of PC glazings create challenges non-existent in glass glazings.

For example, the PC glazings must be protected against abrasion and processes must preferably be developed to incorporate various functional elements within PC glazings. Furthermore, PC can expand and contract more than glass. Accordingly, PC can present issues that do not exist with existing glass glazing technology.

Whenever PC has been implemented on a vehicle, the PC lens has been previously protected with a UV protective varnish. However, the UV protective varnish can degrade over time, leaving the material prone to degradation due to the environment.

Accordingly, there is a need to have a multi-layered plastic panel for use in automotive vehicles wherein the plastic panel has both high impact strength and can withstand various weather conditions without compromising strength.

Figure 2.3 shows a plan view of a section of a glazed plastic panel (75). The plastic panel implements a hybrid film 140. A

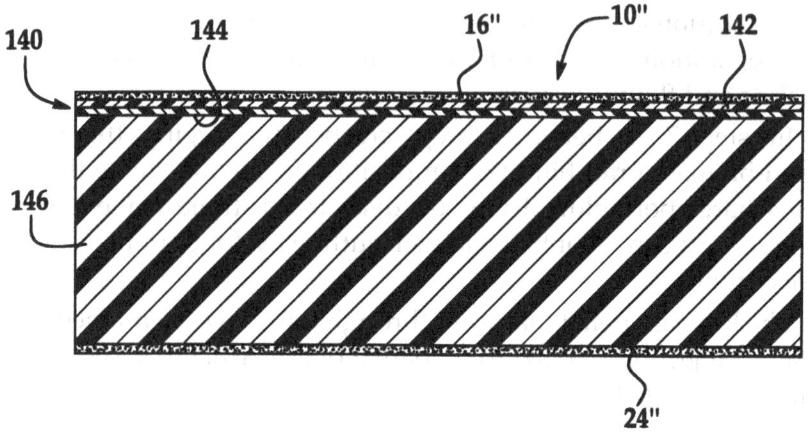

Figure 2.3 Glazed plastic panel (75).

scratch-resistant coating 16" may be applied to the hybrid film 140 and a secondary scratch-resistant coating 24" may be applied to the polycarbonate substrate. The total thickness of the hybrid film 140 may be in the range of approximately 0.25 to 1.0 mm. The hybrid film 140 is formed from two layers. The outermost layer is a poly(methyl methacrylate) (PMMA) film layer 142 with UV absorbers and the inner film layer is a polycarbonate film layer 144. The PMMA film layer 142 may include UV-absorbing material such as but not limited to encapsulated benzophenone-3, avobenzone, octyl methoxycinnamate and diethylamino hydroxybenzoyl hexyl benzoate (75). These UV-absorbing materials are shown in Figure 2.4.

The hybrid film 140 having the combined PMMA film layer 142 and the polycarbonate film layer 144 may be inserted into an injection mold and a polycarbonate layer 146 may be injected behind the hybrid film 140. The hybrid film 140 may be thermoformed and inserted into a mold for complex geometries. Alternatively, the hybrid film 140 may be fed directly into the mold in sheet form for simple geometries. It is further understood that the rigid panel polycarbonate layer 144 may also be injection molded or overmolded onto the hybrid film 140. The polycarbonate substrate 146 may have a thickness in the approximate range of 2.0 to 6.0 *mm*. This construc-

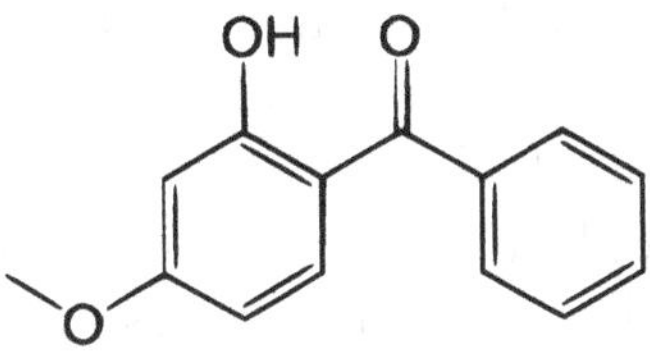

Benzophenone-3

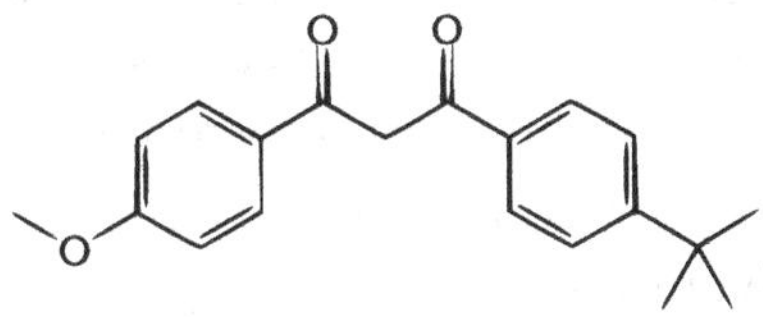

Avobenzone

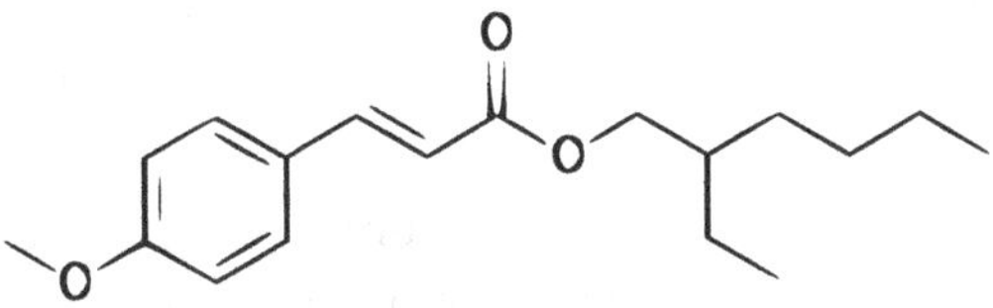

Octyl methoxycinnamate

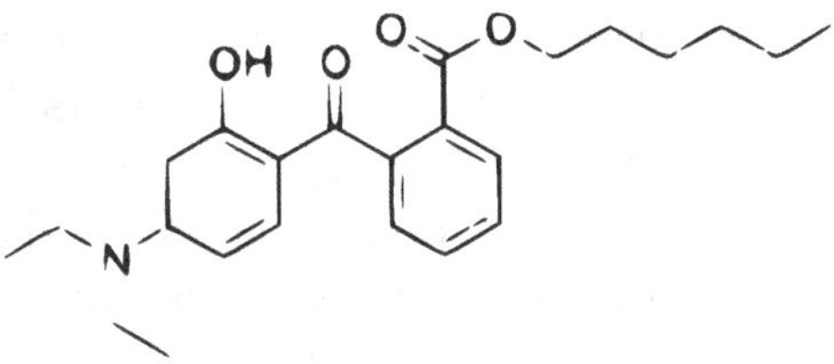

Diethylamino hydroxybenzoyl hexyl benzoate

Figure 2.4 UV-absorbing materials (75).

tion provides for a robust design that can withstand impacts due to the unitary rigid polycarbonate substrate 146 which forms the panel (75).

2.8.4 *Automotive Interiors*

Plastics used for automotive interiors are weakly resistant to deformation by chemicals, corrosion, UV light or heat in terms of the properties thereof, and have surface defects such as scratches, poor surface quality, limited colors, etc., caused upon injection molding, in terms of the outer appearance thereof (76).

In order to alleviate such defects of plastics, a process of applying a solvent-based acrylic lacquer onto the surface of plastics has been conventionally employed.

The solvent-based acrylic lacquer is a two-component solvent-based coating with an acryl or polyester coating and an isocyanate curing agent.

The large amounts of organic solvents in the solvent-based coating may generate volatile organic compounds, thus causing environmental contamination and offensive odors.

An aqueous-based coating has been used instead of the oil coating in the field of plastic parts for automotive interiors. The application of the aqueous coating has received great attention in terms of decreasing the amount of volatile organic compounds. Furthermore, the aqueous coating is advantageous for health and the environment, and as well, and may also overcome cracks or corrosion of the surface of plastics due to the attack of constituents of the solvent-based coating, such as a strong solvent such as xylene or methyl isobutyl ketone.

Intensive research has been done on methods of manufacturing a coating film having a soft feeling similar to natural leather upon contact with a user, using an aqueous coating composition in the field of plastic parts for automotive interiors, which decreases offensive odors resulting from the presence of a non-reactive monomer of an aqueous acrylic resin and a solvent for formation of the coating film (76).

An example for following is an experimental process for such a coating film (76):

First, 60% of an aqueous poly(urethane) (PU) resin, 11.4% of an aqueous PU resin with a hydroxyl group, 10% of a micronized silica dispersant having an average particle size of 10 μm, 1.0% of a PE paraffin wax dispersant having an average particle size of 50 μm, 1.0% of a poly(siloxane) surface controlling agent, 1.5% of a poly-oxy(ethylene) sorbitan monooleate stabilizer, 0.1% of an aqueous solution of lavender-aroma microcapsules, and 10% of ion exchange water were mixed together. Subsequently, to the resulting mixture, 5% of a trifunctional hexamethylenediisocyanate curing agent having a hydrophilic group was added, thus preparing a coating composition (76).

2.8.4.1 *Measurement of the Properties*

In order to evaluate the properties of the aqueous coating composition, the following experiments were conducted. The aqueous coating composition was applied onto a plastic substrate using a sprayer to a thickness of 20 μm based on the thickness of a dry film, dried at 80°C for 20 *min* and then allowed to stand for 24 *h*. The properties of the coating film thus obtained were measured as follows (76):

Outer Appearance. The coating composition was applied onto the plastic substrate, dried at 80°C for 20 *min*, and then observed to determine whether bubbles, impurities, spots or grooves were formed and measured for the degree of smoothness and 60° gloss.

Adhesion. Based on a tape adhesion test according to ASTM D3359 (77), tape was attached to 100 checks of the film, each having a size of 1 *mm* × 1 *mm*, and then removed, after which the number of peeled film checks was counted.

Chemical Resistance. The coating film was repeatedly rubbed 20 times with a soft cloth immersed in a 75% aqueous solution of each of isobutyl alcohol, ethanol, and methanol, after which the results were observed.

Weather Resistance. This test was conducted on a black plate panel at 83±3°C for 500 *h* using a weather-proof tester and a Xenon light source, followed by observing whether problematic defects were actually created.

Scratch Resistance. A scratch measuring unit was mounted to a reciprocally moving device at 0.2 times per *min* under a stroke of 100 *mm*, and was then moved perpendicular to the coating film on the film under a load of 300 *g*, followed by observing whether the film was damaged.

Odor. The completely dried coating film was allowed to stand at room temperature for 7 days, cut to a 10 *cm* × 10 *cm* square, placed into a sealed steel container, and then heated using a heater at 110°C for 5 *h*. Thereafter, the container was allowed to stand at room temperature for 2 *h* and was then opened, and the extent of odor was evaluated.

2.8.5 Automotive Glass Run

An automotive glass run that is attached to an inner periphery of a door frame of an automotive door, guides ascending and descending of a door glass, and is formed of a thermoplastic elastomer, includes: a main body having a substantially U-shaped cross section, the main body including a vehicle-exterior side walls, a vehicle-interior side walls, and a bottom wall; a holding lip that extends from at least one of the vehicle-exterior side walls and the vehicle-interior side wall and abuts the door frame; and a plasticizer-rich layer formed on at least one side of the holding lip abutting the door frame, the plasticizer-rich layer including a large amount of plasticizer component compared to the thermoplastic elastomer forming the holding lip.

The glass run has been conventionally manufactured using vulcanized rubber, such as EPDM, but in consideration of properties close to rubber, a specific gravity lighter than that of a rubber compound, ease of molding, recycling and the like, products formed using a thermoplastic elastomer have been used in recent years. Even in a case of using a thermoplastic elastomer, there is no special change in the configuration of the glass run, and the glass run is basically configured to include the main body having a substantially U-shaped cross section and including the vehicle exterior and interior side wall and the bottom wall (78).

2.8.6 *Milling Burr Control Tape*

An injection molded plastic instrument panel P.S.I.R. hidden door was described (79).

The method consists of applying a stiff tape, with a hardness greater than the hardness of the plastic instrument panel, to a plastic instrument panel surface to be milled to prevent flexing of the surface during milling; milling the surface along a generally widthwise central location of the tape; and removing the tape upon completion of the milling, said method thereby minimizing burrs along edges of the milled surface.

The primary reason for using burrs is the lack of surface support for the plastic during milling as the milling cutter's edge passes. The plastic near the surface (on the side being milled) flexes due to the lack of supporting material above it, thus preventing the complete cutting of this surface material (79).

2.9 Future Trends in Summary

In this chapter, some future trends in the plastics industry were explained. Here, recommendations for future research were also noted.

Also discussed was how plastics have brought enormous societal benefits owing to their versatility, light weight, durability and low costs. However, these properties have come with negative externalities, especially because these persistent materials are leaked into the environment across their entire life cycle.

References

1. Wikipedia contributors, Fourth industrial revolution — Wikipedia, the free encyclopedia, https://en.wikipedia.org/wiki/Fourth_Industrial_Revolution, (2023). [Online; accessed 13-August-2023].

2. S. Echchakoui and N. Barka, Industry 4.0 and its impact in plastics industry: A literature review, *Journal of Industrial Information Integration* 100172 (2020).
3. J.R. Gisbert, C. Palau, M. Uriarte, G. Prieto, J.A. Palazón, M. Esteve, O. López, J. Correas, M.C. Lucas-Estañ, P. Giménez, et al., Integrated system for control and monitoring industrial wireless networks for labor risk prevention, *Journal of Network and Computer Applications* 233–252 (2014).
4. T. Römer and R. Bruder, User centered design of a cyber-physical support solution for assembly processes, *Procedia Manufacturing* 456–463 (2015).
5. G. Schuh, T. Potente, R. Varandani, C. Hausberg, and B. Fränken, Collaboration moves productivity to the next level, *Procedia Cirp* 3–8 (2014).
6. F. Civerchia, S. Bocchino, C. Salvadori, E. Rossi, L. Maggiani, and M. Petracca, Industrial internet of things monitoring solution for advanced predictive maintenance applications, *Journal of Industrial Information Integration* 4–12 (2017).
7. R.Y. Zhong, G.Q. Huang, S. Lan, Q. Dai, X. Chen, and T. Zhang, A big data approach for logistics trajectory discovery from RFID-enabled production data, *International Journal of Production Economics* 260–272 (2015).
8. C. Zioudas, S. Kourmpetis, E. Chamalidou, A.-A. Maniatopoulos, and A. Gazis, An industry 4.0 case study in plastics manufacturing: Bridging the gap between the current state and future trends, *Digital Manufacturing Technology* 78–98 (2022).
9. P.S. Kumar, R. Dinesh, S. Balangnesh, and V. Vijayakumar, Evolution, revolution and future of plastics for an eco-friendly environment, in R. Gupta, D. Deshmukh, A.P. Patil, N.K. Shrivastava, J. Giri, and R. Chadge, (Eds.), *Recent Advances in Material, Manufacturing, and Machine Learning*, Vol. 1 of *Proceedings of 1st International Conference (RAMMML-22*. PSG Colege of Technoogy, CRC Press, (2022).
10. K.S. Pratheesh, R. Dinesh, S. Balaganesh, and V. Vijayakumar, Evolution, revolution and future of plastics for an eco-friendly environment in *Recent Advances in Material, Manufacturing, and Machine Learning* 95–104. CRC Press, (2023).
11. D.M. Mitrano and M. Wagner, A sustainable future for plastics considering material safety and preserved value, *Nature Reviews Materials* 71–73 (2022).
12. A.L. Andrady, *Plastics and Environmental Sustainability*, John Wiley & Sons, (2015).
13. K. Dhali, M. Ghasemlou, F. Daver, P. Cass, and B. Adhikari, A review of nanocellulose as a new material towards environmental sustainability, *Science of the Total Environment* 145871 (2021).

14. A.A. Modi, R. Shahid, M.U. Saeed, and T. Younas, Hemp is the future of plastics, in *E3S Web of Conferences*, Vol. 51 03002. EDP Sciences, (2018).
15. A.T.M.F. Ahmed, M.Z. Islam, M.S. Mahmud, M.E. Sarker, and M.R. Islam, Hemp as a potential raw material toward a sustainable world: A review, *Heliyon* (2022).
16. S. Rachel Jacob, A. Mishra, M. Kumari, K. Bhatt, V. Gupta, and K. Singh, A quick viability test protocol for hemp (Cannabis sativa L.) seeds, *Journal of Natural Fibers* 1281–1286 (2022).
17. T. Karche et al., The application of hemp (Cannabis sativa L.) for a green economy: A review, *Turkish Journal of Botany* 710–723 (2019).
18. M.R.L. Horne, Bast fibres: Hemp cultivation and production, *Handbook of Natural Fibres* 163–196 (2020).
19. S. Amaducci, D. Scordia, F.H. Liu, Q. Zhang, H. Guo, G. Testa, and S.L. Cosentino, Key cultivation techniques for hemp in Europe and China, *Industrial Crops and Products* 2–16 (2015).
20. I. Adesina, A. Bhowmik, H. Sharma, and A. Shahbazi, A review on the current state of knowledge of growing conditions, agronomic soil health practices and utilities of hemp in the United States, *Agriculture* 129 (2020).
21. W. Cierpucha, R. Kozłowski, J. Mańkowski, J. Waśko, and T. Mańkowski, Applicability of flax and hemp as raw materials for production of cotton-like fibres and blended yarns in Poland, *Fibres & Textiles in Eastern Europe* 13–18 (2004).
22. J.P. Manaia, A.T. Manaia, and L. Rodriges, Industrial hemp fibers: An overview, *Fibers* 106 (2019).
23. S. Adanur, *Wellington Sears Handbook of Industrial Textiles*, Routledge, (2017).
24. A.G.D. Schumacher, S. Pequito, and J. Pazour, Industrial hemp fiber: A sustainable and economical alternative to cotton, *Journal of Cleaner Production* 122180 (2020).
25. M. Kostic, B. Pejic, and P. Skundric, Quality of chemically modified hemp fibers, *Bioresource Technology* 94–99 (2008).
26. R.K. Dahal, B. Acharya, and A. Dutta, Mechanical response of the hemp biocarbon-filled hemp-reinforced biopolymer composites, *Journal of Polymer Research* 1–22 (2023).
27. T. Haramina, N. Hadžić, and Z. Keran, Epoxy resin biocomposites reinforced with flax and hemp fibers for marine applications, *Journal of Marine Science and Engineering* 382 (2023).
28. Wikipedia contributors, Circular economy — Wikipedia, the free encyclopedia, https://en.wikipedia.org/wiki/Circular_economy#Engineering_the_Circular_Lifecycle, (2023). [Online; accessed 13-August-2023].

29. D.A. Teigiserova, L. Hamelin, and M. Thomsen, Towards transparent valorization of food surplus, waste and loss: Clarifying definitions, food waste hierarchy, and role in the circular economy, *Science of the Total Environment* 136033 (2020).
30. M. Martins de Oliveira, A. Lago, and G.P. Dal' Magro, Food loss and waste in the context of the circular economy: A systematic review, *Journal of Cleaner Production* 126284 (2021).
31. L. Principato, L. Ruini, M. Guidi, and L. Secondi, Adopting the circular economy approach on food loss and waste: The case of italian pasta production, *Resources, Conservation and Recycling* 82–89 (2019).
32. C. Hanson, B. Lipinski, K. Robertson, D. Dias, I. Gavilan, P. Gréverath, S. Ritter, J. Fonseca, R. VanOtterdijk, T. Timmermans, et al., Food loss and waste accounting and reporting standard (2016).
33. P.C. Slorach, H.K. Jeswani, R. Cuéllar-Franca, and A. Azapagic, Environmental and economic implications of recovering resources from food waste in a circular economy, *Science of the Total Environment* 133516 (2019).
34. M.R. Johansen, T.B. Christensen, T.M. Ramos, and K. Syberg, A review of the plastic value chain from a circular economy perspective, *Journal of Environmental Management* 113975 (2022).
35. J. Payne, P. McKeown, and M.D. Jones, A circular economy approach to plastic waste, *Polymer Degradation and Stability* 170–181 (2019).
36. H.-K. Timken, Circular economy for plastic waste to polyethylene via refinery crude unit, US Patent 11 174 436, assigned to Chevron USA Inc., November 16, 2021.
37. H.-K. Timken, Circular economy for plastic waste to polypropylene via refinery FCC unit, US Patent 11 584 890, assigned to Chevron USA Inc, February 21, 2023.
38. H. Vieyra, J.M. Molina-Romero, J.d.D. Calderón-Nájera, and A. Santana-Díaz, Engineering, recyclable, and biodegradable plastics in the automotive industry: A review, *Polymers* 3412 (2022).
39. J. Diefenderfer, V. Arora, and L. Singer, International energy outlook 2016 liquid fuels, *Doe/Eia-0484* 202–586 (2016).
40. V.G.G. Emissions, 2017 and later model year light-duty vehicle greenhouse gas emissions and corporate average fuel economy standards (2012).
41. M.L. Di Lorenzo and R. Androsch, *Industrial Applications of Poly(lactic acid)*, Vol. 282, Springer, (2018).
42. S.A. Begum, A.V. Rane, and K. Kanny, Applications of compatibilized polymer blends in automobile industry in *Compatibilization of Polymer Blends* 563–593. Elsevier, (2020).
43. A. Adeniyi, O. Agboola, E.R. Sadiku, M. Durowoju, P. Olubambi, A.B. Reddy, I. Ibrahim, and W. Kupolati, Thermoplastic-thermoset nanostructured polymer blends in *Design and Applications of*

Nanostructured Polymer Blends and Nanocomposite Systems 15–38. Elsevier, (2016).

44. A. Patil, A. Patel, and R. Purohit, An overview of polymeric materials for automotive applications, *Materials Today: Proceedings* 3807–3815 (2017).
45. L. Nickels, New innovations in automotive thermoplastics, *Reinforced Plastics* 185–188 (2019).
46. K. Szeteiova, Automotive materials plastics in automotive markets today, *Institute of production technologies, Machine technologies and materials, Faculty of Material Science and Technology in Trnava, Slovak University of technology Bratislava* (2010).
47. L. Kroll, M. Meyer, W. Nendel, and M. Schormair, Highly rigid assembled composite structures with continuous fiber-reinforced thermoplastics for automotive applications, *Procedia Manufacturing* 224–231 (2019).
48. C.-K. Park, C.-D.S. Kan, W.T. Hollowell, and S.I. Hill, Investigation of opportunities for lightweight vehicles using advanced plastics and composites (2012).
49. H. Vieyra, M.A. Aguilar-Méndez, and E. San Martín-Martínez, Study of biodegradation evolution during composting of polyethylene–starch blends using scanning electron microscopy, *Journal of Applied Polymer Science* 845–853 (2013).
50. D. Danso, J. Chow, and W.R. Streit, Plastics: Environmental and biotechnological perspectives on microbial degradation, *Applied and Environmental Microbiology* e01095–19 (2019).
51. H. Vieyra, E. San Martín-Martínez, E. Juárez, U. Figueroa-López, and M.A. Aguilar-Méndez, Biodegradation process of a blend of thermoplastic unripe banana flour-polyethylene under composting: Identification of the biodegrading agent, *Journal of Applied Polymer Science* (2015).
52. R.P. Babu, K. O'connor, and R. Seeram, Current progress on bio-based polymers and their future trends, *Progress in Biomaterials* 1–16 (2013).
53. S. Spierling, E. Knüpffer, H. Behnsen, M. Mudersbach, H. Krieg, S. Springer, S. Albrecht, C. Herrmann, and H.-J. Endres, Bio-based plastics - A review of environmental, social and economic impact assessments, *Journal of Cleaner Production* 476–491 (2018).
54. R. Blackburn, *Sustainable Textiles: Life Cycle and Environmental Impact*, Elsevier, (2009).
55. T. Sen and H.J. Reddy, Application of sisal, bamboo, coir and jute natural composites in structural upgradation, *International Journal of Innovation, Management and Technology* 186 (2011).
56. M. Sain and S. Panthapulakkal, Green fibre thermoplastic composites in *Green Composites* 181–206. Elsevier, (2004).

57. M.S. Fogorasi and I. Barbu, The potential of natural fibres for automotive sector-review, in *IOP Conference Series: Materials Science and Engineering*, Vol. 252 012044. IOP Publishing, (2017).
58. V. Chauhan, T. Kärki, and J. Varis, Review of natural fiber-reinforced engineering plastic composites, their applications in the transportation sector and processing techniques, *Journal of Thermoplastic Composite Materials* 1169–1209 (2022).
59. O. Faruk, A.K. Bledzki, H.-P. Fink, and M. Sain, Biocomposites reinforced with natural fibers: 2000–2010, *Progress in Polymer Science* 1552–1596 (2012).
60. P. Roy, F. Defersha, A. Rodriguez-Uribe, M. Misra, and A.K. Mohanty, Evaluation of the life cycle of an automotive component produced from biocomposite, *Journal of Cleaner Production* 123051 (2020).
61. S. Marur, *Plastics Application Technology for Lightweight Automobiles*, SAE international, (2013).
62. T. Corbière-Nicollier, B.G. Laban, L. Lundquist, Y. Leterrier, J.-A. Månson, and O. Jolliet, Life cycle assessment of biofibres replacing glass fibres as reinforcement in plastics, *Resources, Conservation and Recycling* 267–287 (2001).
63. O.T. Adesina, T. Jamiru, E.R. Sadiku, O.F. Ogunbiyi, and L.W. Beneke, Mechanical evaluation of hybrid natural fibre–reinforced polymeric composites for automotive bumper beam: A review, *The International Journal of Advanced Manufacturing Technology* 1781–1797 (2019).
64. R. Dunne, D. Desai, R. Sadiku, and J. Jayaramudu, A review of natural fibres, their sustainability and automotive applications, *Journal of Reinforced Plastics and Composites* 1041–1050 (2016).
65. T. Volova, A. Boyandin, A. Vasiliev, V. Karpov, S. Prudnikova, O. Mishukova, U. Boyarskikh, M. Filipenko, V. Rudnev, B.B. Xuân, et al., Biodegradation of polyhydroxyalkanoates (PHAs) in tropical coastal waters and identification of PHA-degrading bacteria, *Polymer Degradation and Stability* 2350–2359 (2010).
66. J. Gerrard and M. Kandlikar, Is European end-of-life vehicle legislation living up to expectations? Assessing the impact of the ELV directive on green innovation and vehicle recovery, *Journal of Cleaner Production* 17–27 (2007).
67. F. Carrasco, J. Gámez-Pérez, O.O. Santana, and M.L. Maspoch, Processing of poly (lactic acid)/organomontmorillonite nanocomposites: Microstructure, thermal stability and kinetics of the thermal decomposition, *Chemical Engineering Journal* 451–460 (2011).
68. E. Castro-Aguirre, F. Iniguez-Franco, H. Samsudin, X. Fang, and R. Auras, Poly(lactic acid)–mass production, processing, industrial applications, and end of life, *Advanced Drug Delivery Reviews* 333–366 (2016).

69. M.P. Arrieta, M.D. Samper, J. López, and A. Jiménez, Combined effect of poly (hydroxybutyrate) and plasticizers on polylactic acid properties for film intended for food packaging, *Journal of Polymers and the Environment* 460–470 (2014).
70. B. Wu, P. Xu, W. Yang, M. Hoch, W. Dong, M. Chen, H. Bai, and P. Ma, Super-toughened heat-resistant poly (lactic acid) alloys by tailoring the phase morphology and the crystallization behaviors, *Journal of Polymer Science* 500–509 (2020).
71. OECD, Future prospects for industrial biotechnology, (2011).
72. L.A. Hansen, M. Mcnaughton, A.M. Kowalewski, A.M.-l. Chan, S.M. Gaukler, and C.D. Hathcock, Aser annual site environmental report 2020, *Los Alamos National Lab. (LANL), Los Alamos, NM (United States)* (2022).
73. M. Yasuda, Y. Nakashima, and Y. Ishiguro, Airbag apparatus for automobile, US Patent 7 007 970, assigned to Sanko Gosei Ltd., March 07, 2006.
74. S.H. Ryu, J.H. Park, and S.Y. Ryu, Plastic, for automotive interior parts, manufactured from biomass pellets, and method for manufacturing the plastic, US Patent 20 150 119 503, assigned to SH Global Co. Ltd., April 30, 2015.
75. G. Brooks and G. Benvenuto, Automotive plastic panel, US Patent 10 000 654, assigned to Flex-N-Gate Corp., June 19, 2018.
76. O. Kim, Y. Chun, Y.-H. Cho, K.-M. Jin, J.-B. Ahn, and C.-H. Cho, Aqueous coating composition for plastic automotive interiors, US Patent 7 601 776, assigned to Hyundai Mobis Co. Ltd., Daihan Bee Chemical Co. Ltd., October 13, 2009.
77. Standard test methods for rating adhesion by tape test, ASTM Standard D3359-23, ASTM International, West Conshohocken, PA, (2023). https://www.astm.org/d3359-23.html.
78. H. Otsuka and H. Sugawa, Automotive glass run, US Patent 20 190 232 772, assigned to Toyoda Gosei Co Ltd., August 01, 2019.
79. G.S. Evans, Automotive P.S.I.R. milling burr control tape, US Patent 8 038 168, assigned to Intertec Systems LLC, October 18, 2011.

3

Plastic Waste

3.1 Valorization of Plastic Waste

Nowadays, polymers are the most versatile material (1). They contain certain chemicals and additives such as pigments, concentrates, anti-blockers, light transformers (LTs), UV-stabilizers, etc.

An in-depth analysis has been presented with respect to the recovery, treatment and recycling routes of plastic solid waste (PSW), as well as the main advantages and disadvantages associated with every route (1).

Recovery and recycling of PSW can be categorized by four main routes, i.e., re-extrusion, mechanical, chemical and energy recovery.

Re-extrusion (primary recycling) utilizes scrap plastics by re-introducing the remainder of certain extruded thermoplastics (mainly poly-a-olefins) into heat cycles within a processing line.

When plastic articles are discarded after a number of life cycles, mechanical recycling techniques present themselves as a candidate for utilizing a percentage of the waste as recyclate and/or fillers.

Collectively, all technologies that convert polymers to either monomers (monomer recycling) or petrochemicals (feedstock recycling) are referred to as chemical recycling. The technology behind its success are the depolymerization processes (e.g., thermolysis) that can result in a very profitable and sustainable industrial scheme, providing a high product yield and minimal waste.

Nevertheless, due to their high calorific value and embodied energy, plastics are being incinerated solely or in combination with municipal solid waste (MSW) in many developed countries.

Cement kilns and fluidized beds are the two most common units used to recover energy from PSW or MSW with a high PSW content. It is concluded that tertiary (chemical methods) and quaternary (energy recovery) are robust enough to be investigated and researched in the near future, for they provide a very sustainable solution to the PSW cycle (1).

3.2 Origin of Plastic Waste

3.2.1 Microplastics

The alarming rise of micro-nanoplastics (MNPs) in their environment is currently the biggest threat to biotic and abiotic components around the globe. These pollutants, apart from being formed through fragmentation of larger plastic pieces, are also manufactured for commercial usage. The MNPs enter the agro-ecosystem, wildlife, and the human body through the food chain, ingestion or through inhalation, causing blockage in the blood-brain barrier, lower fertility, and behavioral abnormalities among other problems. Hence, it has become essential to develop novel procedures for remediation of MNPs. Among the numerous existing methods, microbial remediation promises to degrade/recover MNPs via a green route. Since microbial remediation processes mostly depend upon biotic and abiotic factors, such as temperature, pH, oxidative stress, etc., it becomes easy to influence changes in the plastic pollutants. Hence, with the help of recent technologies, a complete degradation/removal of MNPs can be expected by utilizing the respective carbon content as energy sources for growth of microorganisms. In this review, considering the urgent environmental need, the impact of micro-nanoplastic in the ecosystem along with their corresponding degradation mechanisms are discussed. Also, the importance of the various recent research approaches in MNPs remediation is highlighted. Finally, the role of enzyme and membrane technology, nanoparticle technology, and metagenomics in remediation of MNPs is discussed for the first time in detail to bring forth a novel remedy for the environment.

Micro-nanoplastics (MNP) are categorized into two major divisions: primary and secondary MNPs. Primary MNPs originate from household products, cosmetics, and polymeric raw materials of plastic industries such as poly(ethylene) (PE), poly(styrene) (PS) and poly(propylene) (PP). Secondary MNPs are formed due to fragmentation of extensive plastic waste from exposure to abiotic factors such as temperature, UV radiation, microbial degradation, and atmospheric pressure.

3.2.2 *Plastic Waste from Food*

3.2.2.1 *Plastic Packaging Waste*

Concerns about the food delivery industry are growing regarding the management of plastic packaging waste. The amount of plastic packaging from online food delivery services in Korea has been investigated. Also, the environmental effects using life cycle assessments were reported (2).

Furthermore, the environmental impacts of the adoption of multi-use containers, the use of recycled materials, and the increase in recycling rates proposed by the Korean government for efficient plastic waste management were given.

A total of 72.93 *kt* of plastic packaging was consumed by online food delivery in 2020 and PP and poly(ethylene terephthalate) (PET) accounted for 81.48% of the consumed packaging materials.

The adoption of multi-use containers is the most environmentally effective alternative, but its negative impact on terrestrial ecotoxicity is approximately 5 times higher than that of others. Although the other two alternatives are 2–6 times less efficient than adopting multi-use containers, they can still play an important role in plastic waste management.

These results provide empirical information on food packaging waste and insights into the sustainable management of plastics (2).

3.2.2.2 *Microplastics in Human Food*

Plastics are highly resistant to degradation and remain in the environment for a prolonged period of time, resulting in easy access of microplastics into the human food chain (3). The exposure of microplastic to humans is caused by foods of both animal and plant origin, food additives, drinks, and plastic food packaging.

Living organisms can accumulate microplastics in cells and tissues. This results in threats of chronic biological effects and potential health hazards for humans, including body gastrointestinal disorders, immunity, respiratory problem, cancer, infertility, and alteration in chromosomes.

Because of the threat of microplastics on human health, it is essential to ensure food safety as well as to control the use of plastic with a strict regulation of proper management.

Future research is needed on the importance of the core components of microplastics in relation to their exposure to human food, prevention in the human food chain, and biological reactions in the human body.

It has been recommended to consider the presence of microplastics in different foods, as most of the existing research mainly focused on seafood. And it is important to study the mechanism of toxicity with pathways in the human body based on the different types, shapes, and sizes of plastic particles (3).

Micro-nanoplastics in Plant-Derived Food. The contamination of MNPs in plants and plant-derived food has raised concerns about food safety and human health risks. In a review, the sources, contamination pathways, and human exposure risks of MNPs in plant-derived food were examined (4).

Micro-nanoplasticss have been found in fruits, vegetables, staples, and processed plant-derived foods. The major sources of MNPs in plant-derived food include the soil-plant system, air deposition, and food processing and packaging.

Since MNPs can infiltrate plants through root uptake from the soil, via cracks in roots and endocytosis, quantifying human consumption of MNPs from plant- derived food is challenging.

While existing research shows MNPs toxicity in human cells and animals, further investigation is required to determine health risks from MNP exposure through plant diets. In the future, it is crucial to establish detection methods for MNPs, minimize the uptake of soil MNPs by plants, and reinforce related food safety control (4).

Poly(styrene) Nanoplastics. The environmental presence of micro/nanoplastics (MNPLs) is an environmental and human health concern. Such MNPLs can result from the physicochemical/biological degradation of plastic goods (secondary MNPLs) or can result from industrial production at that size for different commercial purposes (primary MNPLs) (5).

Independently of their origin, the toxicological profile of MNPLs can be modulated by their size, as well as by the ability of cells organisms to internalize them. To get more information on these topics the ability of three different sizes of poly(styrene) MNPLs (50 *nm*, 200 *nm*, and 500 *nm*) were determined to produce different biological effects in three different human hematopoietic cell lines (Raji-B, THP-1, and TK6) (5).

Results of the study showed that none of the three sizes were able to induce toxicity (growth ability) in any of the tested cell types. Although transmission electron microscopy and confocal images showed cell internalization in all the cases, their quantification by flow cytometry demonstrated an important uptake by Raji-B and THP-1 cells, in comparison with TK6 cells.

For the first cell types, the uptake was negatively associated with the size. When the loss of mitochondrial membrane potential was determined, dose-related effects were observed for Raji-B and THP-1 cells, but not for TK6 cells. These effects were observed for the three different sizes.

Finally, when oxidative stress induction was evaluated, no clear effects were observed for the different tested combinations. It was concluded that size, biological endpoint, and cell type are aspects modulating the toxicological profile of MNPLs.

3.2.3 *Toxic Products in Plastic Waste*

Incineration of plastic waste in an open field is a major source of air pollution (6). Most of the time, the municipal solid waste containing about 12% plastics is burnt, releasing toxic gases like dioxins, furans, mercury and polychlorinated biphenyls into the atmosphere.

Furthermore, the burning of poly(vinyl chloride) (PVC) liberates hazardous halogens and pollutes the air, the impact of which is climate change.

The toxic substances thus released are posing a threat to vegetation, human and animal health and the environment as a whole.

Poly(styrene) is harmful to the central nervous system. The hazardous brominated compounds act as carcinogens and mutagens.

Dioxins settle on the crops and in our waterways where they eventually enter into our food and hence the body system. These dioxins are the lethal poly(oxypropylene)s and their worst component, 2,3,7,8 tetrachlorodibenzo-*p*-dioxin, commonly known as agent orange, is a toxic compound which causes cancer and neurological damage and disrupts reproductive thyroid and respiratory systems.

Thus, the burning of plastic wastes increases the risk of heart disease, aggravates respiratory ailments such as asthma and emphysema, and cause rashes, nausea or headaches, and damages the nervous system. Hence, a sustainable step towards tomorrow's cleaner and healthier environment needs immediate attention of the environmentalists and scientists.

The hazards of incineration have been presented. Furthermore, open burning of plastics and the effects of plastic in water were described. Also, a possibility of working out strategies to develop alternate procedures of plastic waste management were discussed (6).

3.2.3.1 Chlorinated Plastics

Chlorinated plastics have been widely used as an indispensable thermoplastic owing to their low cost, durability, wide processing adaptability and good overall performance in plenty of end-use applications (7).

An effective pathway to reduce plastic white pollution is to upcycle chlorinated plastics for added value and versatility rather than recycle them.

Upcycling technologies for converting chlorinated waste plastics into additional valuable products and endowing them with added versatility are of interest. Several upcycling strategies for the conversion of chlorinated waste plastics into value-added products were described. These procedures involve a pretreatment to reduce the chlorine content; pyrolysis, carbonization or catalytic cracking; or chemical modifications such as substitution with functional groups and plasticizers, and grafting with other polymers.

In addition, a solvent-based processing is discussed, including solvent extraction, and dissolution, gel casting and solvothermal treatments (7).

3.2.3.2 *Burning of Plastics*

The burning of poly(styrene) or poly(vinyl chloride) is extremely dangerous. Such plastics especially release toxic styrene gas, dioxins, or chlorinated furans when ignited. Interactions between these plastics and other wastes can also lead to toxic emissions mixes (8).

When burned, some plastics are less harmful than other types that yield emissions and residues linked to severe environmental and human health problems. This is shown in Table 3.1.

Table 3.1 Toxicants released upon open burning of various plastics.

Materials	Released Toxicants
Poly(ethylene terephthalate)	Methane, ethane, ethyne, formaldehyde, carbon dioxide, carbon monoxide, polycyclic aromatic hydrocarbons
High-density poly(ethylene)	Olefins, paraffin, aldehydes, and light hydrocarbons, carbon monoxide, polycyclic aromatic hydrocarbons
Poly(vinyl chloride)	Carbon monoxide, dioxins, chlorinated furans, hydrogen chloride, polycyclic aromatic hydrocarbons
Low-density poly(ethylene)	Olefins, paraffin, aldehydes, and light hydrocarbons, carbon monoxide, polycyclic aromatic hydrocarbons
Poly(propylene)	Naphthalene, methyl naphthalene, biphenyl, fluorene, phenanthrene, methyl phenanthrene, anthracene, pyrene, and benzo[a] fluorene, polycyclic aromatic hydrocarbons
Poly(styrene)	Styrene gas, acrolein, hydrogen cyanide, polycyclic aromatic hydrocarbons
Poly(urethane)	Carbon monoxide, hydrogen cyanide, phosgene

3.2.4 *Medical Plastic Waste*

3.2.4.1 *Hazardous Hospital Plastic Waste*

Microwave co-pyrolysis was tested as an approach for the simultaneous reduction and treatment of environmentally hazardous hospital plastic waste (HPW), lignocellulosic (palm kernel shell (PKS)) and triglycerides, i.e., waste vegetable oil (WVO), biowaste as co-feedstock (9).

The co-pyrolysis demonstrated faster heating rate of $16°C\,min^{-1}$ to $43°C\,min^{-1}$ compared to microwave pyrolysis of single feedstock of $9°C\,min^{-1}$ to $17°C\,min^{-1}$. Microwave co-pyrolysis of HPW/WVO performed at 1:1 ratio produced a higher yield (80.5%) of hydrocarbon liquid fuel compared to HPW/PKS (78.2%). The liquid oil possessed a low nitrogen content (< 4%) and was free of sulfur that could reduce the release of hazardous pollutants during its use as fuel in combustion.

In particular, the liquid oil obtained from co-pyrolysis of HPW/WVO has low oxygenated compounds (< 16%), leading to a reduction in the generation of potentially hazardous sludge or problematic acidic tar during oil storage. An insignificant amount of benzene derivatives (< 1%) was also found in the liquid oil, indicating the desirable feature of this pyrolysis approach to suppress the formation of toxic polycyclic aromatic hydrocarbons. Microwave co-pyrolysis of HPW/WVO improved the yield and properties of liquid oil for potential use as a cleaner fuel, whereas the liquid oil from co-pyrolysis of HPW/PKS is applicable in the synthesis of phenolic resin (9).

3.2.4.2 *Alternative Fuel from Medical Plastic Waste*

The production of alternative fuels through medical plastic waste has been reviewed (10). In the recent past, medical plastic waste has been disposed of and incinerated in the dumping yards, which is the main cause of the threat of infection and environmental hazards.

The adoption of a proper waste management system and using the appropriate technology like the 5R's (refuse, reduce, reuse, repurpose, and recycle) may significantly improve the ecosystem.

The 5R's is a comprehensive approach that can be applied either with awareness of stakeholders or enforcement mandate and regulation by the government (10).

A possible route for converting medical plastic waste into drop-in fuel and value-added products to minimize the waste through suitable technology was suggested (10). Here, the pyrolysis technique plays an important role which is more ecologically friendly, effective and produces minimal pollutants.

It has been observed that using a COVID medical waste management (CMWM) technology, 70% to 80% plastic pyrolysis oil (PPO), 10% to 15% bio-char, and gaseous fuel can be extracted.

As per the CMWM, the extracted PPO is a potential feedstock for the CI engine fuel. This review work provides a suitable solution for CMWM and improves the quality of medical infrastructure for sanitation in a sustainable mode (10).

3.3 Waste Accumulation

Synthetic polymers, commonly referred to as plastics, are anthropogenic contaminants that adversely affect the natural ecosystems. The continuous disposal of long lifespan plastics has resulted in the accumulation of plastic waste, leading to significant pollution of both marine and terrestrial habitats (11).

The scientific pursuit to seek environmentally friendly materials from renewable resources has focused on cellulose, the primary reinforcement component of the cell wall of plants, as it is the most abundantly available biopolymer on earth (11).

3.3.1 Reduction of Microplastic Waste Accumulation

Microplastics (MPs) and nanoplastics (NPs) flow into rivers through point pollution and nonpoint pollution sources. Before plastics enter the ecosystem, they need to be filtered and removed. However, the technology for this process is not yet available.

Therefore, the following topics were reviewed: Types and properties of plastics, the after-use condition of MPs and NPs that persist in the environment, the chemical affinity of plastics due to their chemical properties, and bioaccumulation.

Importantly, three key concepts were analyzed: The viable options to manage incidents, where plastics have already entered the environment, the application of bio-inspired technology, and the governmental policies to reduce the generation of plastics through life-cycle management.

Plastics are used in various industries, with the packaging industry (40.5%) using the most in the EU in 2020, followed by the building and construction (20.4%), automotive (8.8%), and electrical and electronics (6.2%) industries. The common plastic material produced in the EU are shown in Table 3.2. Subsequently, special issues of these materials are summarized.

Table 3.2 Plastic material produced in the EU in 2021 (12).

Compound	Amount/[%]
Poly(ethylene)	30.35
Low-density poly(ethylene)	17.45
High-density poly(ethylene)	12.9
Poly(propylene)	9.7
Poly(vinyl chloride)	9.6
Poly(ethylene terephthalate)	8.4
Poly(styrene)	6.1

Plastics are divided into thermosets and thermoplastics according to their properties when heated. Thermosets are rigid and stable, but cannot be reprocessed by softening and molding. On the other hand, thermoplastics are relatively easy to recycle as they can be molded at high temperatures and have an elastic or solid shape at room temperature (13, 14)

Poly(ethylene) (PE) has a variety of uses, feom film, packaging, and electrical insulation to container piping. It can be classified as a low-density poly(ethylene) (LDPE) or high-density poly(ethylene) (HDPE) due to the differences in resistance, toughness, flexibility, and clarity to chemicals (13).

A low-density poly(ethylene) is a solid and flexible polymer characterized by long branches, and an HDPE can be packaged closer to crystallites as polymer chains become more linear (15).

Poly(vinyl chloride) (PVC), the most commonly used plastic resin in medical devices, is chemically non-reactive, rigid, and easily welded and thermoformed (16).

Poly(styrene) (PS) is thermally stable, has high translucency and durability, and is a thermoplastic polymer that can be easily dyed. These characteristics make PS suitable for food storage and transportation and producing packaged products and toys (17).

Poly(propylene) (PP) is a thermoplastic material manufactured by polymerizing propylene molecules in monomer units into long polymer molecules or chains (18, 19). It is lighter in weight than PS due to its lower specific gravity. Furthermore, PP has high rigidity and impact resistance. It is used in slit films, carpets, cast films, and rigid packaging containers.

Poly(ethylene terephthalate) (PET) has low friction and high wear resistance and is resistant to dilute acid, aliphatic and aromatic hydrocarbons, and water at room temperature. It appears white in a semi-crystalline state and transparent in an amorphous state. Amorphous PET has a relatively lower hardness, rigidity, and thermal resistance than crystalline PET (20).

3.3.1.1 Plastic Debris

Plastic debris persists in the environment and can become nanosized; therefore, it is necessary to evaluate the toxicity of plastics according to the size of plastic fragments.

Plastic fragments can be classified into: macroplastics (> 25 *mm*), mesoplastics (5–25 *mm*), MPs (< 5 mm). and NPs (< 100 *nm*). In general, MPs are defined as less than 5 *mm*, with the European Chemicals Agency, the National Oceanic and Atmospheric Administration, and the European Food Safety Authority following the same standard (21, 22).

However, NPs are not clearly defined, because several studies classify plastics with sizes less than 1 μm as NPs (23, 24).

There are two types of MPs: primary MPs and secondary MPs. Primary MPs are produced during the manufacturing stage in sizes that are less than 5 *mm*) and are used in cosmetics and sanitary products. Secondary MPs are introduced into the environment mainly through general waste, improper management of landfills, loss during collection, and wear of tires.

3.4 Conversion of Plastic Waste into Fuel

3.4.1 Apparatus for the Conversion of Plastic Waste into Fuel

Conversion systems for producing fuel in a pyrolysis process have been utilized. Examples of known processes are included in the patents of Miller (25), Srinakruang (26), and Stankevitch (27). All these patents teach processes that use catalyst cracking techniques that depolymerize the hydrocarbon chains and then produce fuel once the plastic is vaporized (28, 29).

Plastic pyrolysis systems currently use large rotating vessel and large kiln apparatuses. A challenge with using large vessels is their inability to fill and discharge continuously as these systems use a batch fill to insert plastic and then heat the product with sealed ends. Although claiming to be a continuous system, they generally utilize multiple units. During the time one unit is being filled, another unit is emptied. Therefore, the process is not truly a continuous process but an overlapping batch process.

A second problem with large vessels of a conventional system includes drawbacks with the sealing mechanisms. Sealing mechanisms expand due to heat and pressure. Due to this expansion, oxygen can leak into the vessels, creating a fire hazard.

A third challenge with the current art is that the startup time of the pyrolysis reaction is prolonged due to the fact that the vessel needs to be started up when it is cold and the oxygen needs to be evacuated before heating. This is an extremely time-intensive process.

Another problem with prior art systems includes vessels that require a cooling-down period before cleaning, for example, in order to access the interior structures.

The extraction of the fuel and the char is also conducted in a batch operation. The removal of the char in prior art is extremely dirty because of the nature of cleaning as the vessel collects char on the wall. This char has to be removed before the next batch is processed.

This method is very time-consuming and expensive due to the low efficiency of combination of heating, cooling, and cleaning.

Other systems use vats that are very similar in nature as the one described above as they must be purged of oxygen and filled with plastic then sealed, then heated, and then cleaned. These machines require large, highly machined surfaces for a good seal. The batch heating and cooling requires large amounts of energy to achieve the pyrolysis process in a timely manner. Manual labor is used in the cleaning process. This is expensive and can be hazardous.

Other plastic systems have been designed only for a special type of plastic. The inability to switch from a certain homogeneous stream to a mixed plastic stream requires a reconfiguration of the equipment.

Different grades, sources, and streams of plastic have different densities and different textures that effect handling parts differently.

Plastic bags, film, and light-density plastics present a problem to standard shredders and feeding mechanisms due to the wrapping effect and bridging that occurs. Wrapping occurs as the plastic bags build up around the cutting shaft, eventually stopping the shredder mechanism. Although knife cutting devices will work to resolve this problem, they will not allow for hard contaminates to pass through the cutting blades. The hard contaminates will damage the cutting edge and even stop the cutter rotation.

An apparatus for processing reusable fuel has been developed (28). This apparatus is shown in Figure 3.1.

It contains a reactor section which employs a supply feeder or bag press feeder 300 to feed plastic material into the reactor 1400 to a heated airlock feeder 1700. The char is dispelled through a char discharge system 1900. Vapors produced within the reactor section may continue throughout the reactor and be collected through an ash separator tower or char separator 1800. The reactor section includes:

1. A continuous material supply assembly,
2. A heated airlock feeder configured to continuously receive and process the material supply received therein,
3. A reactor configured to receive the processed material from the heated airlock feeder, and
4. A vapor refining system configured to process vapor supplied by the reactor.

The apparatus may contain a char disposal system configured to eliminate char from the reactor. The apparatus may also consist of a thermal expansion system configured to allow thermal expansion of

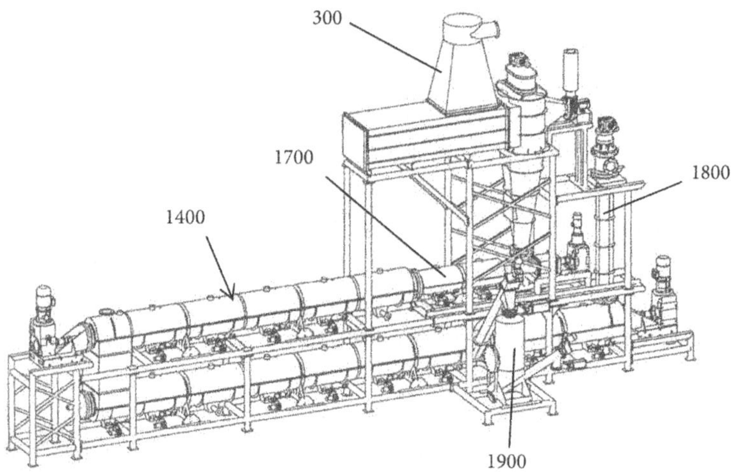

Figure 3.1 Apparatus for processing reusable fuel (28).

300	Press feeder
1400	Reactor
1700	Airlock feeder
1900	Char discharge system
1800	SepaSrator tower or char separator

the reactor. A cooling system may be configured to receive processed fuel from the reactor.

Processing of plastic bags in conventional prior art systems is known to be problematic. For example, due to the characteristically light weight nature of the plastic-type materials, difficulties exist for processing and feeding these materials into an extruder. Plastic bag materials have a tendency to float when moved and form a very large bridge when stored.

Plastic waste material may be shredded and fed, for example, into a reactor (e.g., a pyrolysis reactor). Applying a temperature above 350°C will cause the shredded plastic material to melt and to vaporize. An upstream reactor may utilize back pressure to assist the thermal cracking of carbon chains present in the plastic material, so as to produce a refined fuel or reuseable fuel for subsequent use.

Also, an apparatus assembly may be used to feed the plastic material into a reusable fuel processing unit including a pneumatic feed having a blower as a power source.

The blower may be powered by the heat used to melt and vaporize the shredded plastic. The assembly may also include a screened air outlet, a plurality of press augers, a pinch conveyer system, a destoner hood, a receiver hopper, low and high level indicators, ram, press screw inlet, pneumatic air cylinder, bag press inlet, outlet, and amp monitor.

3.4.2 *Pyrolysis of Poly(ethylene)*

The production of liquid fuels from waste plastics could simultaneously reduce the quantity of waste plastics and provide an alternative fuel source (30). Methods of plastic decomposition include thermal pyrolysis, catalytic cracking, and hydrocracking of plastics to produce short-chain components in compliance with the fuel requirements. The importance of pyrolysis of plastics to create fuel has been explained, the chemistry of plastics and fuels were also described. Furthermore, the mechanisms of thermal pyrolysis and catalytic pyrolysis were described.

A model study of the pyrolysis of PE was performed, looking at the effects of temperature, catalyst, support and bifunctional catalysts. Based on these results, recommendations for future investigations were given (30).

The process to produce high-quality liquid fuel includes melting a solid plastic waste and adding a metal hydride and a supported metal catalyst (30).

The metal hydride can be selected from magnesium hydride (MgH_2), calcium hydride (CaH_2), palladium hydride, beryllium hydride (BeH_2), aluminum hydride (AlH_3), indium hydride (InH_3), lithium aluminum hydride ($LiAlH_4$), sodium aluminum hydride ($NaAlH_4$), sodium borohydride ($NaBH_4$), and a combination of these materials.

The metal catalyst may be selected from CaH_2Al2O3, SiO_2, zeolite, zirconia, MgO, TiO_2, activated carbon, clays, and a combination thereof (30).

3.4.3 Pyrolysis for the Recovery of Aromatic Compounds

Mixed plastic wastes are difficult to mechanically recycle into new products because of their nonhomogeneity. Therefore, pyrolysis and gasification seem to be very effective in the extraction of energy from mixed plastic wastes in the form of oil and gas. In a study, a fraction of mixed plastic wastes was pyrolyzed in a bench-scale plant equipped with a fluidized-bed reactor and a char removal system. The study had two aims (31):

1. The first was to discover the optimum reaction temperature for a high yield of benzene, toluene, and xylene aromatics.
2. The second was to discover the best additive for chlorine removal.

To discover the optimum reaction temperature for the maximum aromatic content, experiments at a temperature range of 660°C to 780°C were carried out. The pyrolysis oils that were obtained from the experiments were composed of aliphatics and mono- and polyaromatic compounds. The reaction temperature had a positive effect on the yield of aromatic compounds. The maximum yield of aromatics was obtained at 719°C, with a value of 18% of the organic product. The pyrolysis oils had almost no metal content.

The metals that were present in the feed material were concentrated in the solid product, the char. It seems that the char removal system, which was composed of a cyclone and a hot filter, almost perfectly removed the char particles.

Besides the pyrolysis oil, a combustible gas was also obtained. The main components of the gas were CH_4, C_2H_4, C_3H_6, C_4H_6, and C_4H_{10}.

The higher heating value of the product gas was about 50 $MJ\,kg^{-1}$.

In order to absorb the hydrogen chloride that was formed from the degradation of the PVC in the mixed plastic wastes, additives (calcium oxide, calcium hydroxide, crushed oyster shells, and rice straw) were added to a fraction of the mixed plastic wastes.

In experiments without any additive, the contents of chlorine in the pyrolysis oils were 350 *ppm* and 500 *ppm*. When the additives, especially calcium oxide, calcium hydroxide, and crushed oyster shells, were applied, the chlorine content of the pyrolysis oils significantly decreased to 50 *ppm*. The results of this study suggest that the use of oyster shells as a HCl absorbent can stimulate the recycling of waste oyster shells (31).

3.4.3.1 Pyrolysis of Poly(styrene)

The catalytic and thermal degradation of waste expandable poly(styrene) (PS) were studied in a semi-batch reactor with continuous flow of nitrogen to achieve greater oil yield and maximize styrene monomer recovery (32).

The effect of temperature, nature of catalyst and its size, reaction time of catalytic pyrolysis, and effect of repyrolysis have also been investigated.

A higher reaction temperature favors the oil yield and also decreases the reaction time with maximum styrene selectivity (76.31%) at 450°C. Among the catalysts that were studied, solid base BaO was found to be the most efficient and increases the styrene selectivity (84.29%) significantly at a reaction temperature of 350°C in comparison to thermal and acid catalytic degradation.

An increase in the size of the catalyst and repyrolysis decrease the oil yield, and styrene production also decreases on repyrolysis (32).

In another study, the pyrolysis of end-of-life PS was studied in a pilot-scale unit consisting of an extruder and a continuous stirred tank reactor (33). Here, the PS was pyrolyzed with continuous feeding at a pressure range from 0.02 to 1.0 *bar*, and a temperature range from 450°C to 600°C, giving primarily styrene, other mono-aromatics, and oligomers.

The styrene yield increased from 36% at 1.0 *bar*, and 450°C to 56% at 0.02 *bar*, and 550°C. Working under a vacuum enhanced the styrene recovery at all corresponding temperature levels. The yield of benzene, toluene, ethylbenzene, and xylene increased from 4% at 450°C and 0.02 *bar* to 17% at 450°C and 1.0 *bar*.

The experimental results have been used in a mathematical model that can explain the combined effect of temperature and pressure on the yield of the primary products (33).

The degradation of PS into styrene, including monomer and dimer, was studied using various kinds of solid acids and bases as catalysts (34). It was found that solid bases were more effective catalysts than solid acids for the degradation of poly(styrene) into styrene. This fact was attributed to differences in the degradation mechanisms of poly(styrene) over solid acids and bases. Among the solid bases employed, BaO was found to be the most effective catalyst, and about 90% of PS was converted into styrene when thermally degraded PS was admitted to BaO powder at 623°K (350°C).

Poly(styrene) films with dispersed BaO powder were then prepared as a model of a recyclable plastic using a twin-roller heated at 393°K. Here, the amount of BaO dispersed was only 1% of the PS films. More than 85% of the PS films with dispersed BaO were converted into styrene by a simple thermal degradation at 623°K (350°C) without assistance of other catalytic compounds.

The films were expanded into PS foams with dispersed BaO using a gas absorption/evolution technique. Styrene could be recovered from these foams by heating them at 673°K (400°C) (34).

3.4.4 Dechlorination of Mixed Plastics Pyrolysis Oils

Waste plastics may contain PVC and/or poly(vinylidene chloride) (PVDC). Through a pyrolysis process, waste plastics can be converted to gas and liquid products. These liquid products may contain paraffins, iso-paraffins, olefins, naphthenes, and aromatic components along with organic chlorides in concentrations of hundreds of *ppm* (35).

However, the liquid products of a pyrolysis process (pyrolysis oils) cannot be used as a feedstock for steam crackers due to feed requirements of very low concentrations of chlorine and limited olefin content.

So, treatment methods were developed for hydrocarbon feedstocks derived from waste plastics to meet certain steam cracker feed requirements (35).

A process for dechlorination of a hydrocarbon stream and/or a hydrocarbon stream precursor was developed. Here, the hydrocarbon stream and/or hydrocarbon stream precursor, a first zeolitic catalyst, and a stripping gas is introduced into a devolatilization extruder to produce an extruder effluent, wherein the hydrocarbon stream and/or hydrocarbon stream precursor contains one or more chloride compounds in an amount of equal to or greater than about 10 *ppm* chloride, based on the total weight of the hydrocarbon stream.

A configuration of a dechlorination system for processing mixed plastic waste is shown in Figure 3.2.

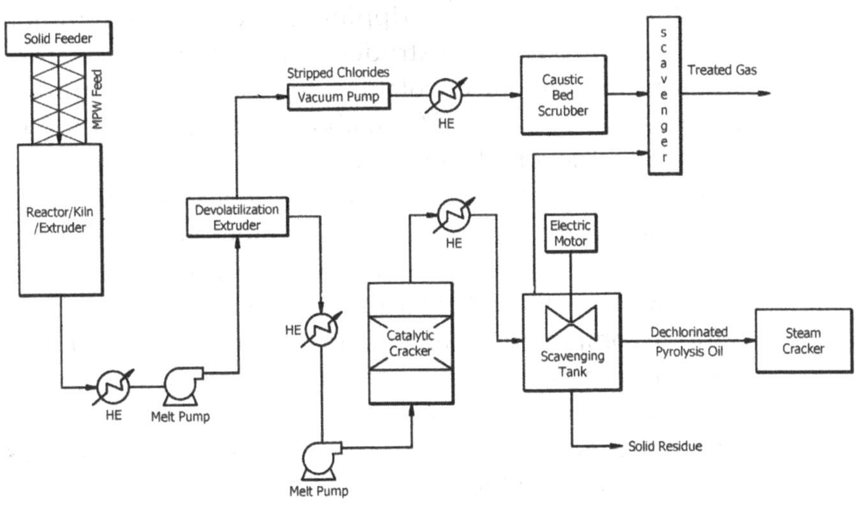

Figure 3.2 Apparatus for dechlorination (35).

The dechlorination system may additionally hydrogenate olefins to provide for a dechlorinated feed that meets requirements for introduction to a steam cracker. The system includes a feeder, a devolatilization extruder, a chloride stripping unit, a scavenging vessel or tank, and a steam cracker (35). Waste plastics can be loaded into or fed to the devolatilization extruder via hydrocarbon stream precursor.

Mixed waste plastics can contain chlorinated plastics (e.g., chlorinated PE, PVC, PVDC), non-chlorinated plastics (e.g., polyolefins, PE, PP, PET), poly(butylene terephthalate), PS, and the like, or mixtures thereof. Waste plastics can also include used tires.

Also described is a process for processing mixed plastic waste by (35):

1. Introducing a mixed plastic waste to a thermal cracking reactor to produce a hydrocarbon stream in a liquid phase and a gaseous stream, wherein the thermal cracking reactor is characterized by a temperature of from about 300°C to about 400°C, and wherein the hydrocarbon stream includes one or more chloride compounds in an amount of equal to or greater than about 10 *ppm* chloride, based on the total weight of the hydrocarbon stream.
2. Introducing at least a portion of the hydrocarbon stream, a zeolitic catalyst, and a stripping gas to a devolatilization extruder to produce an extruder effluent and a spent stripping gas, wherein the devolatilization extruder is characterized by a temperature of from about 150°C to about 450°C, wherein the devolatilization extruder is characterized by a pressure of from about 1 *atm* absolute to about 10 *torr* to 14 *torr*, wherein the devolatilization extruder is characterized by a residence time of from about 0.1 *min* to about 1 *h* or more, wherein the extruder effluent includes one or more chloride compounds in an amount of less than the chloride amount in the hydrocarbon stream, wherein a viscosity of the extruder effluent is lower than a viscosity of the hydrocarbon stream, and wherein the spent stripping gas includes at least a portion of the stripping gas and at least a portion of one or more chloride compounds of the hydrocarbon stream.
3. Introducing at least a portion of the extruder effluent to a catalytic cracker to produce a hydrocarbon product stream in a liquid phase and a gaseous product stream, wherein a viscosity of the hydrocarbon product stream is lower than a viscosity of the extruder effluent, and wherein the catalytic cracker is characterized by a temperature of from about 350°C to about 730°C.
4. Dechlorinating at least a portion of the hydrocarbon prod-

uct stream to yield a polished hydrocarbon stream and a chlorine-containing gaseous stream, wherein dechlorinating occurs at a temperature of from about 25°C to about 225°C and includes agitating, stirring, magnetic stirring, passing through a fixed or fluidized adsorption bed of adsorbents, or combinations thereof; and wherein the polished hydrocarbon stream includes one or more chloride compounds in an amount of less than about 3 *ppm* chloride, based on a total weight of the polished hydrocarbon stream.

5. Feeding at least a portion of the polished hydrocarbon stream to a steam cracker to yield a product, wherein the product includes ethylene, propylene, butene, butadiene, aromatic compounds, or combinations thereof.
6. Contacting at least a portion of the spent stripping gas from the devolatilization extruder, at least a portion of the gaseous stream from the thermal cracking reactor, at least a portion of the gaseous product stream from the catalytic cracker, or combinations thereof, with a caustic solution to remove a portion of one or more chloride compounds and to produce a treated spent stripping gas.
7. Contacting at least a portion of the treated spent stripping gas and/or at least a portion of the chlorine-containing gaseous stream with a chloride adsorber to remove at least a portion of the chloride and to produce a treated gaseous stream.
8. Separating olefin gases as a product from the treated gaseous stream to provide a separated treated gas stream.
9. Recycling at least a portion of the separated treated gas stream as stripping gas to the devolatilization extruder.

Another similar process for processing mixed plastic waste was described. This process uses the following steps (35):

1. Introducing a mixed plastic waste to a thermal cracking reactor to produce a hydrocarbon stream in a liquid phase and a gaseous stream, wherein the thermal cracking reactor is characterized by a temperature of from about 350°C to about 730°C, and wherein the hydrocarbon stream contains one or more chloride compounds in an amount equal to or

greater than about 10 *ppm* chloride, based on the total weight of the hydrocarbon stream.

2. Introducing at least a portion of the hydrocarbon stream, a zeolitic catalyst, and a stripping gas to a devolatilization extruder to produce an extruder effluent and a spent stripping gas, wherein the devolatilization extruder is characterized by a temperature of from about 150°C to about 450°C, wherein the devolatilization extruder is characterized by a pressure of from about 1 *atm* absolute to about 10 *tor* to 14 *tor*, wherein the devolatilization extruder is characterized by a residence time of from about 0.1 *min* to about 1 *h* or more, wherein the extruder effluent includes one or more chloride compounds in an amount of less than the chloride amount in the hydrocarbon stream, wherein a viscosity of the extruder effluent is lower than a viscosity of the hydrocarbon stream, and wherein the spent stripping gas includes at least a portion of the stripping gas and at least a portion of one or more chloride compounds of the hydrocarbon stream.
3. Dechlorinating at least a portion of the extruder effluent to yield a polished hydrocarbon stream and a chlorine-containing gaseous stream, wherein dechlorinating occurs at a temperature of from about 25°C to about 225°C and under agitating, stirring, magnetic stirring, passing through a fixed or fluidized adsorption bed of adsorbents, or combinations thereof; and wherein the polished hydrocarbon stream includes one or more chloride compounds in an amount of less than about 3 *ppm* chloride, based on the total weight of the polished hydrocarbon stream,
4. Feeding at least a portion of the polished hydrocarbon stream to a steam cracker to yield a high-value product, wherein the high-value product contains ethylene, propylene, butene, butadiene, aromatic compounds, or combinations thereof.
5. Contacting at least a portion of the spent stripping gas from the devolatilization extruder and/or at least a portion of the gaseous stream from the thermal cracking reactor with a caustic solution to remove a portion of one or more chloride compounds from the spent stripping gas gaseous stream and to produce a treated spent stripping gas.

6. Contacting at least a portion of the treated spent stripping gas and/or at least a portion of the chlorine-containing gaseous stream with a chloride adsorber to remove at least a portion of the chloride from the treated spent stripping gas and/or at least a portion of chloride from the chlorine-containing gaseous stream and to produce a treated gaseous stream.
7. Separating olefin gases as a high-value product from the treated gaseous stream to provide a separated treated gas stream.
8. Recycling at least a portion of the treated gaseous stream and/or at least a portion of the separated treated gas stream as stripping gas to the devolatilization extruder.

3.4.5 *Hydrogen-Rich Fuel Gas*

Experiments were performed to maximize the production of hydrogen gas from refuse plastic fuel by pyrolysis and steam gasification. The processes were conducted in a laboratory-scale reactor (36).

Gasification processes were carried out using 10 *g* refuse plastic fuel samples at temperatures reaching from 700°C to 1000°C with and without steam. The effect of the amount of steam (0 to 0.25 $g\ min^{-1}$) for refuse plastic fuel steam gasification was also studied. Furthermore, the effect of K_2CO_3 as a catalyst on these processes was tested.

It could be shown that the hydrogen gas yield increased with temperature. With respect to the gas composition, the hydrogen content increased mainly at the expense of other gaseous compounds, which highlights the major extension of secondary cracking reactions in the gaseous fraction at higher temperatures (36).

3.4.6 *Mixed Plastic Waste*

Different types of plastic wastes have caused serious environmental problems. Co-pyrolysis and separation techniques are being widely used because of the complex components of plastic wastes. Thermogravimetric analysis of mixed plastics with different compositions (Polypropylene/Polyvinyl chloride and

Polypropylene/Polyvinyl chloride/poly(styrene) mixtures) was carried out with three heating rates (37).

Aiming at the complex pyrolysis reaction process of mixed plastics, a model-free method and a model-fitting method were combined to obtain the complete kinetics, which makes the results more reliable. The activation energy of the mixture pyrolysis process was carried out by the model-free method, and two stages of pyrolysis mechanism functions were summarized in the model-fitting method.

A verification test proved that the mechanism function of this research can be reliable to express the pyrolysis reaction process. This research can provide support for the optimization of pyrolysis process parameters and the design of pyrolysis reactors for chlorine-containing mixed plastics (37).

3.5 Future Trends in Summary

The valorization of plastic waste is an important feature in this chapter. Nowadays, polymers are considered the most versatile material. They contain certain chemicals and additives, such as, pigments, concentrates, anti-blockers, LTs, UV-stabilizers, etc.

In this chapter, an in-depth analysis was presented with respect to the recovery, treatment and recycling routes of plastic solid waste, as well as the main advantages and disadvantages associated with every route.

The recovery and recycling of plastic solid waste can be categorized by four main routes, i.e., re-extrusion, mechanical, chemical and energy recovery.

References

1. S.M. Al-Salem, P. Lettieri, and J. Baeyens, The valorization of plastic solid waste (PSW) by primary to quaternary routes: From re-use to energy and chemicals, *Progress in Energy and Combustion Science* 103–129 (2010).
2. Y. Jang, K.N. Kim, and J. Woo, Post-consumer plastic packaging waste from online food delivery services in South Korea, *Waste Management* 177–186 (2023).

3. A. Al Mamun, T.A.E. Prasetya, I.R. Dewi, and M. Ahmad, Microplastics in human food chains: Food becoming a threat to health safety, *Science of the Total Environment* 159834 (2023).
4. Y. Liang, X. Cao, A. Mo, J. Jiang, Y. Zhang, W. Gao, and D. He, Micro (nano) plastics in plant-derived food: Source, contamination pathways and human exposure risks, *TrAC Trends in Analytical Chemistry* 117138 (2023).
5. A. Tavakolpournegari, B. Annangi, A. Villacorta, G. Banaei, J. Martin, S. Pastor, R. Marcos, and A. Hernández, Hazard assessment of different-sized polystyrene nanoplastics in hematopoietic human cell lines, *Chemosphere* 138360 (2023).
6. R. Verma, K.S. Vinoda, M. Papireddy, and A.N.S. Gowda, Toxic pollutants from plastic waste-A review, *Procedia Environmental Sciences* 701–708 (2016).
7. S. Xu, Z. Han, K. Yuan, P. Qin, W. Zhao, T. Lin, T. Zhou, and F. Huang, Upcycling chlorinated waste plastics, *Nature Reviews Methods Primers* 44 (2023).
8. G. Pathak, M. Nichter, A. Hardon, E. Moyer, A. Latkar, J. Simbaya, D. Pakasi, E. Taqueban, and J. Love, Plastic pollution and the open burning of plastic wastes, *Global Environmental Change* 102648 (2023).
9. W.A.W. Mahari, S. Awang, N.A.Z. Zahariman, W. Peng, M. Man, Y.-K. Park, J. Lee, C. Sonne, and S.S. Lam, Microwave co-pyrolysis for simultaneous disposal of environmentally hazardous hospital plastic waste, lignocellulosic, and triglyceride biowaste, *Journal of Hazardous Materials* 127096 (2022).
10. A. Kumar, H.S. Pali, and M. Kumar, A comprehensive review on the production of alternative fuel through medical plastic waste, *Sustainable Energy Technologies and Assessments* 102924 (2023).
11. K. Dhali, M. Ghasemlou, F. Daver, P. Cass, and B. Adhikari, A review of nanocellulose as a new material towards environmental sustainability, *Science of the Total Environment* 145871 (2021).
12. H. Lee, J.E. Shim, I.H. Park, K.S. Choo, and M.-K. Yeo, Physical and biomimetic treatment methods to reduce microplastic waste accumulation, *Molecular & Cellular Toxicology* 13–25 (2023).
13. J. Oliveira, A. Belchior, V.D. da Silva, A. Rotter, Ž. Petrovski, P.L. Almeida, N.D. Lourenço, and S.P. Gaudêncio, Marine environmental plastic pollution: Mitigation by microorganism degradation and recycling valorization, *Frontiers in Marine Science* 567126 (2020).
14. O.D. Agboola and N.U. Benson, Physisorption and chemisorption mechanisms influencing micro (nano) plastics-organic chemical contaminants interactions: A review, *Frontiers in Environmental Science* 167 (2021).

15. J.L. Jordan, D.T. Casem, J.M. Bradley, A.K. Dwivedi, E.N. Brown, and C.W. Jordan, Mechanical properties of low density polyethylene, *Journal of Dynamic Behavior of Materials* 411–420 (2016).
16. L.W. McKeen, Plastics used in medical devices in *Handbook of Polymer Applications in Medicine and Medical Devices* 21–53. Elsevier, (2014).
17. K. Kik, B. Bukowska, and P. Sicińska, Polystyrene nanoparticles: Sources, occurrence in the environment, distribution in tissues, accumulation and toxicity to various organisms, *Environmental Pollution* 114297 (2020).
18. W. Kissel, J.H. Han, and J. Meyer, *Handbook of Polypropylene and Polypropylene Composites*, Polypropylene: Structure, Properties, Manufacturing Processes, and Applications, Marcel Dekker, Inc., New York, NY, (2003).
19. Q.T. Shubhra, A.M. Alam, and M.A. Quaiyyum, Mechanical properties of polypropylene composites: A review, *Journal of Thermoplastic Composite Materials* 362–391 (2013).
20. L.N. Ji, Study on preparation process and properties of polyethylene terephthalate (PET), in *Applied Mechanics and Materials*, Vol. 312 406–410. Trans Tech Publ, (2013).
21. Panel on Contaminants in the Food Chain (CONTAM), Presence of microplastics and nanoplastics in food, with particular focus on seafood, *EFSA Journal* e04501 (2016).
22. L. Anagnosti, A. Varvaresou, P. Pavlou, E. Protopapa, and V. Carayanni, Worldwide actions against plastic pollution from microbeads and microplastics in cosmetics focusing on European policies. Has the issue been handled effectively?, *Marine Pollution Bulletin* 111883 (2021).
23. H.K. Imhof, R. Sigl, E. Brauer, S. Feyl, P. Giesemann, S. Klink, K. Leupolz, M.G. Löder, L.A. Löschel, J. Missun, et al., Spatial and temporal variation of macro-, meso-and microplastic abundance on a remote coral island of the Maldives, Indian Ocean, *Marine Pollution Bulletin* 340–347 (2017).
24. B.S.S. Ramasamy and S. Palanisamy, A review on occurrence, characteristics, toxicology and treatment of nanoplastic waste in the environment, *Environmental Science and Pollution Research* 43258–43273 (2021).
25. S.J. Miller, System for producing transportation fuels from waste plastic and biomass, US Patent 8 696 994, assigned to Chevron USA Inc, April 15, 2014.
26. J. Srinakruang, Process for producing fuel from plastic waste material by using dolomite catalyst, US Patent 8 344 195, assigned to Jumluck Srinakruang, January 01, 2013.

27. V. Stankevitch, Process for the conversion of waste plastics to produce hydrocarbon oils, US Patent US6 534 689, assigned to Pyrocat Ltd, March 18, 2003.
28. A.F. Tenore, N. Canosa, and O. Oluwaseun, System and process for converting waste plastic into fuel, US Patent 20 180 010 049, assigned to Golden Renewable Energy LLC, January 11, 2018.
29. A.F. Tenore, D. Fowler, O. Oluwaseun, and A.N. Tenore, System and process for converting waste plastic into fuel, US Patent 20 200 369 964, assigned to Braven Environmental LLC, November 26, 2020.
30. C. Huang, A. Gujar, and M. Rodgers, Methods of producing liquid hydrocarbon fuels from solid plastic wastes, US Patent 9 200 207, assigned to University of Central Florida Research Foundation Inc UCFRF, December 01, 2015.
31. M.-H. Cho, S.-H. Jung, and J.-S. Kim, Pyrolysis of mixed plastic wastes for the recovery of benzene, toluene, and xylene (BTX) aromatics in a fluidized bed and chlorine removal by applying various additives, *Energy Fuels* 1389–1395 (2010).
32. J.J. Park, K. Park, J.-S. Kim, S. Maken, H. Song, H. Shin, J.-W. Park, and M.-J. Choi, Characterization of styrene recovery from the pyrolysis of waste expandable polystyrene, *Energy Fuels* 1576–1582 (2003).
33. A. Zayoud, H.D. Thi, M. Kusenberg, A. Eschenbacher, U. Kresovic, N. Alderweireldt, M. Djokic, and K.M. Van Geem, Pyrolysis of end-of-life polystyrene in a pilot-scale reactor: Maximizing styrene production, *Waste Management* 85–95 (2022).
34. H. Ukei, T. Hirose, S. Horikawa, Y. Takai, M. Taka, N. Azuma, and A. Ueno, Catalytic degradation of polystyrene into styrene and a design of recyclable polystyrene with dispersed catalysts, *Catalysis Today* 67–75 (2000).
35. M. Javeed, A. Stanislaus, A.A. Hakeem, G. Koripelly, R. Narayanaswamy, and K.K. Ramamurthy, Dechlorination of mixed plastics pyrolysis oils using devolatilization extrusion and chloride scavengers, US Patent 10 829 696, assigned to Sabic Global Technologies BV, November 10, 2020.
36. S. Lee, S.-K. Yoo, J. Lee, and J.-W. Park, Hydrogen-rich fuel gas production from refuse plastic fuel pyrolysis and steam gasification, *The Journal of Material Cycles and Waste Management* 191 (2009).
37. F. Yin, Q. Zhuang, T. Chang, C. Zhang, H. Sun, Q. Sun, C. Wang, and L. Li, Study on pyrolysis characteristics and kinetics of mixed plastic waste, *Journal of Material Cycles and Waste Management* 1984–1994 (2021).

4

Plastic Pollution in the Environment

Plastics are ubiquitous in the environment. Extensive studies have focused on analytical methods, source, abundance, transport, fate, and degradation of plastics in the environment and threats to natural surroundings, wildlife or even human health (1).

However, characteristics of plastic pollution, which are critical to understand this emerging problem, have remained unknown.

The major characteristics of plastic pollution in the environment to enhance our present understanding of this issue have been reviewed (1).

4.1 Ingestion of Macroplastics by Odontocetes of the Greek Seas

Plastic pollution is an omnipresent problem that threatens marine animals through ingestion and entanglement (2). Marine mammals are no exception to this rule, but their interaction with plastic remains understudied in the Mediterranean Sea.

In a study, the stomach contents of 34 individuals from seven odontocete species stranded in Greece were analyzed. Macroplastic (> 5 *mm*) was found in the stomachs of nine individuals from four species (harbor porpoise *Phocoena phocoena*, Risso's dolphin *Grampus griseus*, Cuvier's beaked whale *Ziphius cavirostris* and sperm whale *Physeter macrocephalus*), with the highest frequency of occurrence in sperm whales (60%). Gastric blockage from plastic was presumably lethal in three cases, with plastic bags being the most common finding (46%).

Plastic ingestion is of particular conservation concern for the endangered Mediterranean sperm whales. A regular examination of stranded cetaceans with a standardized protocol is critical for allowing spatiotemporal comparisons within and across species.

Anthropogenic marine debris or litter, is defined as anything solid, made or processed by people, and deliberately discarded into or unintentionally lost in the coastal and oceanic environment.

Although debris can include wood, metal, glass, rubber, clothing or paper, plastic is its most prevalent and widespread component (2). So, plastic makes up 60–80% of debris found in the marine environment (3). About 5 to 13 million tons of plastic are discarded into the ocean annually (4). These materials are finding their way from coastal areas to mid-ocean gyres and abyssal depths (5).

Plastic pollution is one of the most pervasive, ubiquitous and long-lasting anthropogenic impacts in the marine environment (6, 7). These impacts are not restricted by national borders and may not be easily reversible.

Two fundamental types of interactions between organisms and debris occur: entanglement and ingestion. Lethal effects of entanglement include drowning, while sub-lethal ones involve skin lesions, compromised feeding, limited predator avoidance capabilities, and reduced reproductive capacity and growth, which eventually lead to reduced fitness (8, 9).

Lethal and sub-lethal effects of debris ingestion cannot be easily identified but available evidence suggests that ingestion can cause ulcerations, perforations and obstruction of the digestive tract, followed by disrupted digestion, feeling of satiation, starvation and general debilitation (10, 11).

The ingestion of debris in sea birds (12) and sea turtles (13) has been documented (2).

4.2 Greenhouse Gas Emissions Associated with Plastics Consumption

Plastics are a ubiquitous part of modern life, but their accumulation in landfills and leakage into the natural environment is now recognized as a global environmental crisis. In response to this massive waste challenge, the scientific community is vigorously pursuing new technologies that can enable displacement of fossil-based carbon feedstocks with bio-based or waste-sourced feedstocks, greener manufacturing of plastics, and development of new approaches that will enable recycling of waste plastics to a much greater extent (14).

A baseline supply of chain energy and greenhouse gas emissions data for the consumption of major plastics in the United States has been shown. This is shown in Figure 4.1.

All of the polymers reported have a global consumption greater than 1 *MMT* per year.

For example, overall energy consumption in the United States is some $3{,}400 \times 10^{15}$ J/year.

Such estimates could serve as a useful benchmark for evaluating new circular economy-focused technologies.

4.3 Sustainability of Plastic Types

Reducing the carbon intensity of plastics production by sourcing sustainable feedstocks while simultaneously enabling effective polymer recycling represents a potential transformation of 21st century manufacturing.

To evaluate technologies that could enable such changes, it is imperative to compare the sustainability of bio-based and/or circular plastic flows to those of incumbent manufacturing paradigms. To that end, we estimate the supply chain energy requirements and greenhouse gas (GHG) associated with US-based plastics consumption. Major commodity polymers, each of which has a global consumption of at least 1 *MMT* per year, account for an estimated annual 3.2 quadrillion BTUTs (quads) of energy and 104 *MMT* CO_2e of greenhouse gas emissions (GHGEs) in the US alone. This study serves as a foundation for comparing the supply chain energy requirements and GHGEs of today's plastics manufacturing

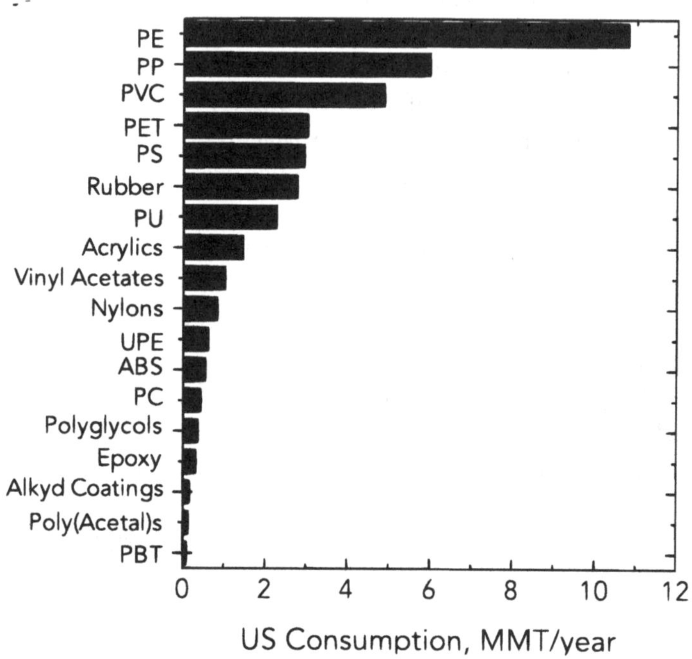

Figure 4.1 US consumption of major plastics (14).

PE	Poly(ethylene)
PP	Poly(propylene)
PVC	Poly(vinyl chloride)
PET	Poly(ethylene terephthalate)
PU	Poly(urethane)
PS	Poly(styrene)
UPE	Unsaturated polyester
ABS	Acrylonitrile-butadiene-styrene
PC	Poly(carbonate)
PBT	Poly(butylene terephthalate)

Poly(styrene)s (PSs) includes poly(styrene) copolymers. Acrylics include poly(acrylic acid) and poly(methyl methacrylate). Vinyl acetates include ethylene vinyl acetate and poly(vinyl alcohol).

to tomorrow's disruptive technologies, to inform the development of bio-based plastics and the circular economy for synthetic polymers (14).

The production of commodity organic chemicals today is both primarily sourced from and powered by fossil carbon resources. Toward decarbonization of this key global economic sector, it is important to understand the contributions to energy usage and GHGEs along the petrochemical manufacturing supply chain, which can inform judicious policy development and impactful technology options to improve or reimagine existing manufacturing practices.

The flow of materials throughout industry can be used to estimate the supply chain energy and GHGEs for 51 organic petrochemicals and 6 intermediates that are globally produced at a capacity of at least 1 million metric tons (*MMT*) per year. This analysis focuses on supply chains in the United States, from which industrial data are readily sourced to obtain accurate energy and GHGE estimates.

Analysis for each chemical includes contributions from sourcing chemical feedstocks, electricity use, and fuel usage for transportation and manufacturing.

This analysis indicates that process fuel, which is primarily used for heating, dominates GHGEs in all cases except for chlorochemicals, where electricity is used extensively for the chloroalkali process and results in a large electricity GHGE contribution ranging from 7 to 54% of total GHGEs.

The contribution of electricity to GHGEs ranges from 6% to 63%. This represents the decarbonization potential in the transition toward renewable electricity with existing manufacturing processes (14).

4.4 Plastic Industry in China

The plastic industry is a high carbon emission industry in China (15). Previous studies mainly focused on the flow and stock of PS, poly(vinyl chloride) (PVC), and acrylonitrile-butadiene-styrene (acrylonitrile-butadiene-styrene (ABS)) plastics. However, there is limited detailed information on the GHGEs of PS, PVC, and ABS in China and the key paths of carbon peak in 2030. Therefore, the authors established the GHG accounting model of PS, PVC, and ABS based on dynamic material flow analysis and life-cycle assessment, and further analyzed the carbon peak path and strategy in 2030.

The results show that the main source of GHG is the production and manufacturing stages. Notably, the key paths of GHGEs are the manufacturing process of PS STP C (standard temperature and pressure) and foam plastics, PVC film and STP C, and ABS STP C. Moreover, the cumulative GHGEs of these plastics reached 230.34, 677.43, 143.12 Tg CO_2e in 2007-2017, respectively.

From 2007 to 2017, the trends of GHGEs from the PS, PVC, and ABS had their characteristics. PVC had the largest GHGEs. Different from PVC and ABS, GHGEs from PS showed a trend of slow rise and then slow decline in this period. According to the current trend, the life-cycle GHGEs of PS, PVC, and ABS would not reach the carbon peak in 2030. Under some mitigation strategies, the carbon peak would be achieved before 2030. This study clarifies the main sources of life-cycle GHGEs of these plastics, and reveals the path and strategy of carbon peak in 2030.

Human activities are responsible for almost all of the increase in GHGEs in the atmosphere, which are the main cause of global warming (16). In 2016, 175 countries signed the Paris Agreement, the main purpose of which is to control the global average temperature rise within 2°C and strive to achieve zero GHGEs in the second half of the 21st century. China is one of the world's largest carbon-emitting countries. In 2017, total carbon emission in China accounted for about 28.27% of the global total and had not yet reached its peak (International Energy Agency, 2019). In 2020, China proposed striving to achieve the goal of carbon peak by 2030 and carbon neutrality by 2060.

According to China's plan, the energy efficiency of key industries would reach the international advanced level by 2030, and the energy consumption per unit of gross domestic product (GDP) would be greatly reduced. The carbon emissions per unit of GDP would decrease by more than 65% compared with 2005; the consumption proportion of non-fossil energy would reach about 25%; and the carbon emissions would reach the peak and decrease steadily.

It was suggested that China will stop adding to the global warming problem by 2060 (17). The proportion of non-fossil energy consumption would reach more than 80%, and the goal of carbon neutrality would be successfully achieved by 2060. However, the energy demand of China is increasing with the fast-growing economy and China relies heavily on coal for its electricity, which will make it difficult to achieve carbon neutrality.

Therefore, the energy demand and supply planning in the future for China should be optimized (17). A Long-range Energy Alternatives Planning system was used to forecast the end-use energy demand of China. Here, a new mixed-integer linear programming model was developed to optimize the energy structure, infrastructure projects and exploitation schemes under the constraint of greenhouse gas emissions (17). Furthermore, economically feasible and sustainable energy planning was also obtained.

It was shown that the electrification rate, green hydrogen demand and other energy demands of China will increase substantially. The coal production capacity will be concentrated, the integration of oil production and refining will be gradually realized, and the production of non-fossil power will increase considerably. The total greenhouse gas emissions of the energy consumption in China between 2017 and 2060 will be approximately 262,783 million tons of CO_2 equivalents (17).

The plastic industry is an industry with high energy consumption and carbon emissions. Plastic products are expected to contribute 15% to global GHGEs by 2050 (18).

Greenhouse gas is emitted in the plastics life-cycle, including the production (P), processing and manufacturing (PM), recycling, landfilling, and incineration (19).

Plastics can be widely used in various industries such as packaging and transportation. In recent years, the yield of plastics has been amazing, and 8.3 billion tonnes of plastics were produced worldwide (20).

The huge production of plastics has brought with it a lot of GHGEs. In the United States, the emissions from the exploitation and transportation of fossil fuels (mainly shale gas) related to plastics production were 9.5-10.5 million tonnes of CO_2 equivalent in 2015 (19).

The research showed that plastic consumption in the United States contributed to the GHGEs of 104 million metric tons CO_2 equivalent (14). By 2030, the net GHGEs from the incineration of plastic packaging waste will increase to 84 million tonnes.

The processes of production and incineration for plastics contributed more than 850 million tonnes of GHG, equivalent to the GHGEs of 189 coal-fired power plants (500 *MW*) in 2019.

4.5 Carbon Footprint

At present, greenhouse gas emissions (GHGEs) associated with plastics have attracted extensive attention.

In recent years, there has been an increase in both consumer and legislative pressure on the packaging sector to reduce the environmental impact of its products, which are often only single-use items. Using data from a plastics manufacturer, a cradle-to-grave study was conducted for trays produced from recycled poly(ethylene terephthalate) (PET), calculating their product carbon footprint and analyzing how various parameters affect the carbon footprint (21).

A model based on a spreadsheet analysis has been developed. This model allows the product carbon footprint to be determined using production batch data.

It was found that the cradle-to-grave carbon footprint of 1 *kg* of recycled polyethylene terephthalate trays containing 85% recycled content was 1.538 *kg* CO_2e. The raw material, manufacturing, secondary packaging, transport and end-of-life stages each contributed 45%, 38%, 5%, 3% and 9% of the total life-cycle of greenhouse gases respectively. The recycled content of raw material was found to have a significant effect on product carbon footprint: A 24% decrease in tray carbon footprint could be obtained by manufacturing trays from 100% recycled content, compared to the current recycled content level of 85%.

A reduction in tray weight was found to give almost an equivalent proportionate reduction in carbon footprint, with 20% and 30% tray weight reductions resulting in product carbon footprint reductions of 18.7% and 28% respectively. Transport was found to only contribute a minor amount of the greenhouse gases of 3% and so improving the transport efficiency has only a small effect on the carbon footprint.

The effect of end-of-life treatment was also found to be relatively small. The worst case scenario of no recycling taking place in the end-of-life stage results in the carbon footprint of the trays increasing by 2.7%, while increasing the recycling rate from 23.7% to 32% and 50% results in the carbon footprint decreasing by 1% and 3% respectively.

In both the extrusion and thermoforming processes, the specific manufacturing carbon footprints arising from consumption of electricity, chilled water energy and compressed air were found to decrease logarithmically with production speed. The greatest reductions in the carbon footprint of recycled polyethylene terephthalate trays can be achieved in the raw material and manufacturing life-cycle stages. The proportion of recycled raw material should be maximized while extrusion and thermoforming process speeds should be optimized as significant manufacturing energy reductions can be attained when the speeds of both processes are increased.

Tray lightweighting should be implemented to as great an extent as possible without compromising tray structural integrity while high recycling rates in the end-of-life stage should continue to be targeted (21).

4.6 Global Greenhouse Gas Emission from Both Traditional Plastics and Bioplastics

Zheng and Suh analyzed global greenhouse gas emissions (GHGEs) from both traditional plastics and bioplastics throughout their life-cycle (22). They found that the global GHG emissions of traditional plastics are projected to grow to 650 million tonnes of CO_2e by 2050.

Gomes *et al.* carried out a systematic review on the life-cycle assessment (LCA) studies of PET packaging materials (23). Brizga analyzed the potential environmental consequences of replacing petrochemical packaging plastics with bioplastics (in terms of GHGEs, land and water footprints) based on a review of LCA studies and calculations to assess the footprint of such alternatives (24).

Su *et al.* quantified the environmental impact of CO_2e of raw material production of express packaging, box manufacturing, and waste management (25). The study showed that there was a great opportunity to reduce the consumption and impact of packaging materials.

Also, a bottom-up model was constructed, which studied the pathways to net-zero GHGEs of 90% global plastics. The results showed that net-zero GHGEs could be achieved by combining the efficient utilization of carbon dioxide (CO_2) and biomass.

Alsabri and Al-Ghamdi analyzed the LCA and carbon footprint of the production process of PVC, poly(ethylene), and poly(propylene) plastics to find the environmental impact leading to global warming (26).

It was found that PVC had higher energy consumption and carbon emissions than other plastics. In conclusion, although these studies involve the carbon footprint of plastics, there are few studies on life-cycle GHG emissions from the most commonly used PS, PVC, and ABS plastics in China, especially on the carbon peak path and strategy of these plastics in 2030. In addition, most of the previous studies on plastic carbon footprint are based on LCA, which can assess the level of environmental impacts throughout its life cycle (from material generation to waste treatment), which cannot be revealed by material flow analysis (MFA). However, the detailed mass flow and balance in the MFA system can provide inventory or activity level for LCA, which will solve the difficulty of obtaining the activity level required for GHGEs accounting in the past. Therefore, this study explores the GHGEs of the plastics life cycle in combination with LCA (15).

4.7 Future Trends in Summary

Plastics are ubiquitous in the environment. Extensive studies have focused on the analytical methods, source, abundance, transport, fate and degradation of plastics in the environment and their threat to natural surroundings, wildlife or even human health.

However, characteristics of plastic pollution, which are critical to understand this emerging problem, have remained unknown.

The major characteristics of plastic pollution in the environment to enhance our present understanding of this issue were reviewed in this chapter.

References

1. P. Li, X. Wang, M. Su, X. Zou, L. Duan, and H. Zhang, *Bulletin of Environmental Contamination and Toxicology* 577–584 (2021).
2. P. Alexiadou, I. Foskolos, and A. Frantzis, Ingestion of macroplastics by odontocetes of the Greek Seas, Eastern Mediterranean: Often deadly!, *Marine Pollution Bulletin* 67–75 (2019).
3. J.G. Derraik, The pollution of the marine environment by plastic debris: A review, *Marine Pollution Bulletin* 842–852 (2002).
4. J.R. Jambeck, R. Geyer, C. Wilcox, T.R. Siegler, M. Perryman, A. Andrady, R. Narayan, and K.L. Law, Plastic waste inputs from land into the ocean, *Science* 768–771 (2015).
5. S. Chiba, H. Saito, R. Fletcher, T. Yogi, M. Kayo, S. Miyagi, M. Ogido, and K. Fujikura, Human footprint in the abyss: 30 year records of deep-sea plastic debris, *Marine Policy* 204–212 (2018).
6. C.J. Moore, Synthetic polymers in the marine environment: A rapidly increasing, long-term threat, *Environmental Research* 131–139 (2008).
7. D.K. Barnes, F. Galgani, R.C. Thompson, and M. Barlaz, Accumulation and fragmentation of plastic debris in global environments, *Philosophical Transactions of the Royal Society B: Biological Sciences* 1985–1998 (2009).
8. D.W. Laist, Overview of the biological effects of lost and discarded plastic debris in the marine environment, *Marine Pollution Bulletin* .
9. M.R. Gregory, The hazards of persistent marine pollution: Drift plastics and conservation islands, *Journal of the Royal Society of New Zealand* 83–100 (1991).
10. J.K. Jacobsen, L. Massey, and F. Gulland, Fatal ingestion of floating net debris by two sperm whales (Physeter macrocephalus), *Marine Pollution Bulletin* 765–767 (2010).
11. M.L. Brandão, K.M. Braga, and J.L. Luque, Marine debris ingestion by Magellanic penguins, Spheniscus magellanicus (Aves: Sphenisciformes), from the Brazilian coastal zone, *Marine Pollution Bulletin* 2246–2249 (2011).
12. M. Codina-García, T. Militão, J. Moreno, and J. González-Solís, Plastic debris in Mediterranean seabirds, *Marine Pollution Bulletin* 220–226 (2013).
13. Q.A. Schuyler, C. Wilcox, K.A. Townsend, K.R. Wedemeyer-Strombel, G. Balazs, E. van Sebille, and B.D. Hardesty, Risk analysis reveals global hotspots for marine debris ingestion by sea turtles, *Global Change Biology* 567–576 (2016).
14. S.R. Nicholson, N.A. Rorrer, A.C. Carpenter, and G.T. Beckham, Manufacturing energy and greenhouse gas emissions associated with plastics consumption, *Joule* 673–686 (2021).

15. J. Chu, Y. Zhou, Y. Cai, X. Wang, C. Li, and Q. Liu, Life-cycle greenhouse gas emissions and the associated carbon-peak strategies for PS, PVC, and ABS plastics in China, *Resources, Conservation and Recycling* 106295 (2022).
16. S. Solomon, D. Qin, M. Manning, K. Averyt, and M. Marquis, *Climate change 2007-the physical science basis: Working group I contribution to the fourth assessment report of the IPCC*, Vol. 4, Cambridge University Press, (2007).
17. S. Qiu, T. Lei, J. Wu, and S. Bi, Energy demand and supply planning of China through 2060, *Energy* 121193 (2021).
18. E. Macarthur and H. Heading, How the circular economy tackles climate change, *Ellen MacArthur Found* 1–71 (2019).
19. L.U. Hamilton, *Economics of Ecosystem Services: Valuation of Open Space and Residential Development*, Villanova University, (2019).
20. R. Geyer, J.R. Jambeck, and K.L. Law, Production, use, and fate of all plastics ever made, *Science Advances* e1700782 (2017).
21. A. Dormer, D.P. Finn, P. Ward, and J. Cullen, Carbon footprint analysis in plastics manufacturing, *Journal of Cleaner Production* 133–141 (2013).
22. J. Zheng and S. Suh, Strategies to reduce the global carbon footprint of plastics, *Nature Climate Change* 374–378 (2019).
23. V. Gomes, N.N. Barros, and R.C. Ruschel, Building information modelling for whole-building LCA: BIM4LCA, in *IOP Conference Series: Earth and Environmental Science*, Vol. 290 012044. IOP Publishing, (2019).
24. J. Brizga, K. Hubacek, and K. Feng, The unintended side effects of bioplastics: Carbon, land, and water footprints, *One Earth* 45–53 (2020).
25. Y. Su, H. Duan, Z. Wang, G. Song, P. Kang, and D. Chen, Characterizing the environmental impact of packaging materials for express delivery via life cycle assessment, *Journal of Cleaner Production* 122961 (2020).
26. A. Alsabri and S.G. Al-Ghamdi, Carbon footprint and embodied energy of PVC, PE, and PP piping: Perspective on environmental performance, *Energy Reports* 364–370 (2020).

5

Recycling

5.1 The Frontier of Plastics Recycling

5.1.1 Waste as a Resource for High-Value Applications

Following the circular economy hierarchy of reduce and reuse, recycling is the third layer to close the loops for materials and decouple their value from consumption (1).

Polymeric materials, i.e., plastics, are in principle well suited for recycling, as they can be reprocessed with relatively low energy input as a material, cleaved back into their monomers or converted back to feedstock.

Today, these approaches still fall short in quantitatively diverting waste towards reuse. Here, the industry challenges were described for recycling back into high-value applications of polymers, and the portfolio of existing and emerging technology solutions.

Sustainable design of products and polymers, recycling technologies, appropriate business models, and enabling technologies converge at the frontier of plastics recycling, for a transformation of the industry towards circularity and greenhouse gas emission neutrality.

5.2 Recycling Technologies

Various methods of plastics recycling have been summarized in Table 5.1.

Table 5.1 Methods of plastics recycling (2).

Plastic waste type	Mechanical recycling	Chemical recycling	Energy recovery
Single plastic waste	++	+	+
Mixed plastic waste	+	++	++
Mixed plastic waste plus paper, etc.	-	-	-

(-): Not applicable
(+): Suitable
(++): Preferred

The compatibility of various plastic materials is shown in Table 5.2.

There are still a lot of other compatibility tables available in the literature (3).

5.2.1 *Maintaining the Polymer Structure*

5.2.1.1 Mechanical Recycling

For thermoplastic polymers, which by volume constitute more than 75% of all plastics used worldwide, melting and re-extrusion is the most straightforward way of recycling.

The reprocessing technique uses the same machinery and workflows as for the conversion of virgin plastics.

Such mechanical recycling (4) provides the bulk of all recycled polymeric materials today. Being a physical process, the polymer structure remains (largely) unchanged. However, the thermal stress of reprocessing adds to any damage to the molecular chains that the material already has experienced due to aging owing to its initial processing, oxidative stress during use, UV exposure, or other factors. The quality of mechanical recycling outputs is directly dependent on the purity and material quality of the input for remelting. A key problem is components that are non-miscible in the majority matrix in the melt (5, 6). They can create uncontrolled heterogeneities in the material, weaken the mechanical performance, and can make processes like film blowing impossible, or deteriorate optical appearance.

Table 5.2 Compatibility of different plastic materials (7).

Polymers	Additives											
	PE	PVC	PS	PC	PP	PA	POM	SAN	ABS	PBT	PET	PMMA
PE	CO	NC	NC	NC	NC	NC	NC	NC	NC	NC	NC	NCL
PVC	NC	CO	NC	NC	NC	NC	NC	CO	LC	NC	NC	CO
PS	NC	NC	CO	NC	NC	NC	NC	NC	NC	NC	NC	NC
PC	NC	NC	NC	CO	NC	NC	NC	CO	CO	CO	CO	CO
PP	SC	NC	NC	NC	CO	NC	NC	NC	NC	NC	NC	NC
PA	NC	NC	SC	NC	NC	CO	NC	NC	NC	SC	SC	NC
POM	NC	NC	NC	NC	NC	NC	CO	NC	NC	SC	NC	NC
SAN	NC	NC	NC	CO	NC	NC	NC	CO	NC	NC	NC	CO
ABS	NC	LC	NC	CO	NC	SC	NC	NC	SC	NC	NC	CO
PBT	NC	NC	NC	CO	NC	NC	NC	NC	SC	CO	NC	NC
PET	NC	NC	SC	CO	NC	SC	NC	NC	SC	NC	CO	NC
PMMA	NC	CO	SC	CO	NC	NC	SC	CO	CO	NC	NC	CO

CO: Compatible
NC: Not compatible
LC: Limited compatible
SC: Compatible in small quantities

Polymer Types
Poly(ethylene) (PE)
Poly(vinyl chloride) (PVC)
Poly(styrene) (PS)
Poly(carbonate) (PC)
Poly(propylene) (PP)
Poly(amide) (PA)
Poly(oxymethylene) (POM)
Styrene acrylonitrile copolymer (SAN)
Acrylonitrile-butadiene-styrene (ABS)
Poly(butylene terephthalate) (PBT)
Poly(ethylene terephthalate) (PET)
Poly(methyl methacrylate) (PMMA)

Hazardous materials, legacy additives that have been phased out or are banned for specific use (8), odor-bearing compounds or chemically active substances may catalyze the polymer degradation or ingredients imparting undesired color may act as contaminants. Therefore, technically rather undemanding applications are still typical for many recyclates. These include street posts, garbage cans, or wood substitutes in construction. High-value applications, on the other hand, target closed-loop scenarios (1).

Chain Extenders for Mechanical Recycling. In some polymers, such as poly(ethylene terephthalate) (PET), bifunctional reactive molecules can be added as chain extenders for mechanical recycling (5, 9).

The molecules allow a reaction between end-groups of oligomers or polymer chains. This results in a buildup of the molecular weight, and restores or improves the mechanical properties.

Specific tools for reprocessing such components are compatibility tables for plastic materials. In these tables, impurities of immiscible polymer components can be found (3).

Compatibility tables can list a lot possibilities of compatibility of commonly employed plastics; however, most tables do not provide further information about how compatible they are (3).

The limitations of these compatibility tables were evaluated in a study. Also, some background is provided here for developing a new compatibility table. The compatibility concept is evaluated both under the view of the design and the polymer engineering and different polymer mixtures and blends are gathered from the literature. Finally, a new compatibility table, based on the compatibility concept adopted by materials engineering, was presented for some commonly employed polymers (3).

5.2.2 *Chemical Safety Aspects*

Food packaging facilitates the storage, handling, transport, and preservation of food (10). Despite these beneficial aspects, food packaging raises environmental concerns due to its high production volume, often short usage time, and problems related to waste management and littering.

Chemical safety aspects of recycled food packaging were investigated. Recycling may increase the levels of potentially hazardous chemicals in the packaging and after migration in the food. Since the exposure to certain chemicals migrating from food packaging has been associated with chronic diseases, it is of high importance to assess the safety of recycled packaging products.

Typical migrants from all types of recycled food packaging materials have been listed. Also, some approaches to reduce chemical contamination are described (10).

General properties of food packaging materials with respect to their recyclability are listed in Table 5.3.

Table 5.3 General properties of food packaging materials (10).

Materials	Recycling feasibility	Comments
Thermoplastics	Yes	PET beverage bottles
Thermosets	No	
Plastic multilayers	Yes	
Paper & board	Yes	No contact with food
Metals	Yes	
Glass	Yes	

5.2.3 *Migration of Contaminants*

5.2.3.1 *Model Contaminants*

The migration of several model contaminants from PET bottles into an Iranian yogurt drink has been investigated (11).

Poly(ethylene terephthalate) (PET) is produced from ethylene glycol and either terephthalic acid or dimethyl terephthalate. It is commonly used as packaging material for drinking water, mineral water, carbonated beverages and edible oils.

The strength and permeability properties of PET, its resistance to chemicals and its high degree of transparency are the main factors that make it more applicable. Hazardous compounds should be avoided as the bottles are manufactured by a commercial plant that produces food contact bottles. However, PET is known to contain small amounts of low-molecular-weight compounds ranging from monomers to pentamers. The level of these compounds depends on the type of PET.

Migration from packaging materials into product generally is concerned with minor constituents that influence the quality of the contained product by sensory or toxicological hazards.

There has been extensive research performed on the use of food contact PET materials. Investigations have been carried out in order to establish the type and concentration of compounds related to post-consumer PET recycling (12–16).

The amount obtained by Soxhlet extraction represents the maximum level of migration of contaminants.

Such an amount may never be attained even under long-term storage conditions. The results of gas chromatography analysis of an extract solution showed that among the three selected contaminants, di(2-ethylhexyl) phthalate had the highest concentration in a PET bottle. The results are shown in Table 5.4.

Di(2-ethylhexyl) adipate and di(2-ethylhexyl) phthalate are shown in Figure 5.1.

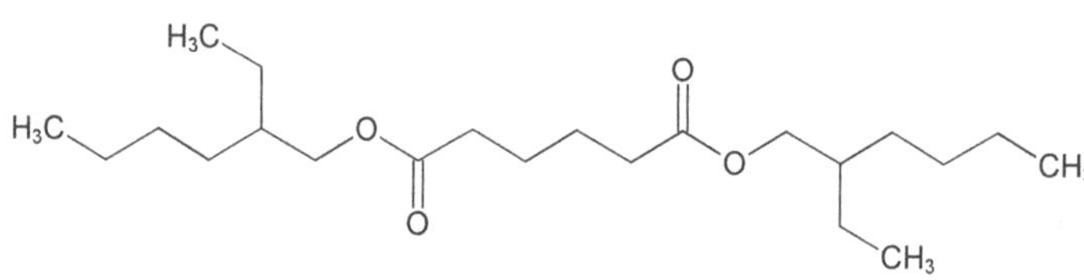

Di(2-ethylhexyl) adipate

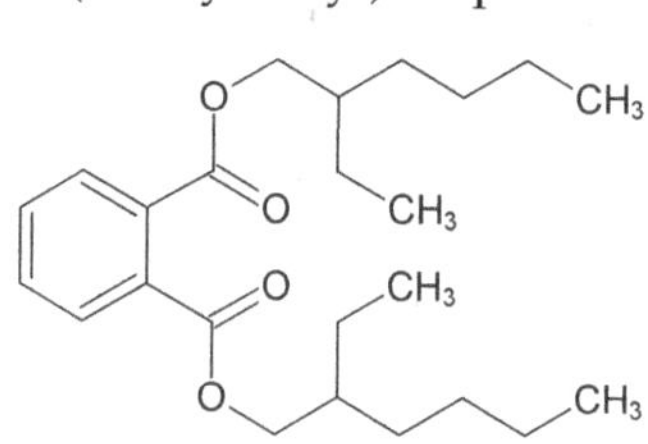

Di(2-ethylhexyl) phthalate

Figure 5.1 Di(2-ethylhexyl) adipate and di(2-ethylhexyl) phthalate.

Table 5.4 Amount of migrants (11).

Compound	Amount/[$\mu g/g$ PET]
Ethylene glycol	21
Di(2-ethylhexyl) phthalate	670
Di(2-ethylhexyl) adipate	510

Furthermore, the migration of di(2-ethylhexyl) adipate, di(2-ethylhexyl) phthalate and ethylene glycol from PET bottles into the stimulant at 25±1°C as a function of time is shown in Table 5.5.

Table 5.5 Migration as function of time (11).

	Compounds		
	Di(2-ethylhexyl) adipate	Di(2-ethylhexyl) phthalate	Ethylene glycol
Time/[d]	Amount of migration/[$mg\,l^{-1}$]		
0	0.0	0.0	0.0
9	1.1	1.4	1.4
16	1.7	1.8	2.2
25	2.0	2.1	2.5
52	2.1	2.3	2.8
66	2.3	2.5	2.9
115	2.3	2.5	3.0

The equilibrium level of migration is not just related to the primary concentration of the monomers in the PET bottles, but the solubility of these compounds in solution is an inevitable factor affecting this phenomenon. Ethylene glycol was found in the highest concentration in food simulant at both storage temperatures of 25 and 45°C. Possibly the higher solubility of ethylene glycol in 3% acetic acid is the main reason for a higher migration of this monomer compared to the other compounds.

5.2.4 *Migration of Plasticizers*

The migration of di(2-ethylhexyl) adipate and acetyltributyl citrate plasticizers from a food-grade poly(vinyl chloride) (PVC) film into a sweetened sesame paste Halawa tehineh was investigated (17).

A loss of 54.7% di(2-ethylhexyl) adipate from the PVC film was observed. This value is slightly higher than the limit of 3 $mg\,dm^{-2}$ of a film surface set by the European Union for di(2-ethylhexyl) adipate.

5.2.5 *Migration of Aluminum and Silicon*

Samples of PET/clay nanocomposite were prepared by melt blending PET and Cloisite 20A nanoparticles (18, 19). A stretch blow-molding machine was used to produce bottles from neat PET and PET nanocomposite. Tests were performed on the migration of aluminum and silicon from PET nanocomposite bottles into acidic food simulant.

The samples were stored at room temperature of 25°C and 45°C for time durations ranging from 7 *d* to 90 *d*. A specific surface of sheets (prepared from PET/clay nanocomposite) immersed in acidic food simulant, and two-sided migration of Al and Si was investigated. According to X-ray diffraction analysis, the nanoclays show intercalated structure in the PET matrix.

Transmission electron microscopy and atomic force microscopy micrographs displayed both intercalation and exfoliation morphology for PET/clay nanocomposites. Inductively coupled plasma was used to quantify amounts of Al and Si that had migrated into the acidic food simulant. It was observed that the migration process is dependent on storage time and temperature, and the molar ratio of aluminum and silicon in the acidic aqueous solution (Al/Si)aq to the ratio in the solid phase of prepared nanocomposites (Al/Si)solid was about 23% higher in the samples stored at 45°C (19).

5.2.6 *Delamination*

Multi-material multilayers may include adhesives for bonding interlayers. So, a recycling process may include delamination steps (20).

A post-consumer liquid packaging board delamination through bonding damage was investigated (21).

Acetone was the solvent used to delaminate food packaging containing a PE/Aluminum/PET composition (22).

The application of physical processes to delaminate multilayers is unusual due to the bonding's strength. However, basic structures could be separated by shredding and washing (23).

An innovative technology has been developed to establish a low-energy material recycling process to delaminate multi-material multilayer plastic packaging (MMPP) (24). This process includes a shredding step followed by a microemulsion with a surfactant that promotes the separation of the layers.

This technique can delaminate structures such as PE/Aluminum, PP/Aluminum, and PE/PET by minimizing interlayers forces.

5.2.7 *Separation*

A raw mixture of plastics usually includes various kinds of waste plastics. This makes the separation an important process that should be carried out prior to recycling. However, the separation of mixed plastics encounters many problems due to the characteristics of plastics. So, this is one of the most problematic processes in the management system of plastic waste (25).

5.2.7.1 *Wet Separation Techniques*

Flotation Separation. As wetting agents for wet separation, sodium ligninsulfonate, tannic acid, diethylhexyl sodium sulfosuccinate and saponin can be used (26). As plastics for testing, PVC, poly(carbonate) (PC), poly(oxymethylene) (POM), and poly(phenylene ether) (PPE) were used.

The efficient flotation separation amongst these naturally hydrophobic polymers could be attributed to the selective reduction in their hydrophobicity as a consequence of the adsorption of the surfactants. The relative order of floatability was found to be PPE > POM > PC > PVC.

This order is consistent with the critical surface tension values determined for all these solids except PPE (26).

5.2.8 *Mechanical Recycling*

During mechanical recycling, polymer backbones can become partially degraded because heating can lead to breakage of intramolecular bonds, resulting in a lower molecular weight distribution and changed mechanical and optical properties of the recycled polymer, e.g., increased brittleness and opacity (27, 28).

Additionally, plastic polymers may already have been damaged by external influences during use or waste management, e.g., contact with acidic substances, UV radiation, and extreme temperatures. These inherent properties of plastics have led to their classification as nonpermanent materials, although they are highly resistant to complete degradation in the environment.

Some types of plastic additives, e.g., antioxidants and stabilizers, form intended reaction products, thereby losing their original function. To compensate for such unwanted changes, mechanical recycling processes require the addition of virgin polymers and/or further additives (29).

5.2.9 *Selective Dissolution*

Selective dissolution techniques are solvent-based methods to separate layers of MMPP. These methods take advantage of the different solubilities of polymers, where a specific solvent dissolves a targeted component and removes materials in steps (30).

For thermoplastic polymers, such as poly(ethylene) (PE), poly(propylene) (PP), poly(styrene) (PS), PET, and PVC, the method of dissolution is an important technique for recycling systems (31). The choice of the solvent is an essential step of this practice, while a range of solvents may allow the recovery of diverse layers (32).

Along the process, the target polymer is dissolved, and other polymers remain at a solid phase. Then, the solvent is used to target another polymer, or else, another solvent can be used (33).

Also, specific components of the multilayer structure can be selectively dissolved for further recovery, with low contamination rates (34). The recycling of beverage cartons using solvent-based techniques was also evaluated (35).

In summary, these methods produce high-quality end products comparable to virgin materials (34) and show that solvent-based technologies are a realistic recycling approach (32).

5.2.10 *Dissolution/Reprecipitation Technique*

The application of the dissolution/reprecipitation technique in the recycling of polymers from waste plastic packaging materials has been investigated (36).

Here, the type of polymer in each packaging was identified with Fourier transform infrared spectroscopy (FTIR). Also, the experimental conditions of the recycling process, i.e., type of solvent/nonsolvent, initial polymer concentration and dissolution temperature, were optimized using model polymers.

The dissolution/reprecipitation technique was applied in the recycling of a number of plastic materials based on PE, PP, PS, PET, and PVC.

The recovery of the polymer was measured and possible structural changes during the recycling procedure were assessed by FTIR.

Furthermore, the potential recycling-based degradation of the polymer was investigated by measuring the thermal properties, such as melting point, crystallinity and glass transition temperature, of the polymer before and after recycling, using differential scanning calorimetry, their molecular properties (average molecular weight) using viscosimetry, as well as their mechanical tensile properties (36).

High recoveries were found in most samples. The properties of the recycled materials were not substantially different from the original materials. However, a slight degradation was observed in a few samples (36).

A method for separating polymers from a physically commingled solid mixture containing a plurality of polymers can be done by the following steps (37):

1. Dissolving at least one first polymer of the mixture in a selected solvent at a first temperature to form a first solution and a first remaining solid component which contains at least two polymers of the mixture which are insoluble in the selected solvent at the first temperature but soluble in the selected solvent at higher temperatures;
2. Separating the first solution from the first remaining solid component;
3. Separating the at least one first polymer from the selected solvent of the first solution;

4. Dissolving at least one second polymer from the first remaining solid component in the selected solvent at a second temperature to form a second solution and a second remaining solid component which contains at least one third polymer of the mixture which is insoluble in the selected solvent at the second temperature but soluble in the selected solvent at a third higher temperature;
5. Separating the second solution from the second remaining solid component;
6. Separating the at least one second polymer from the selected solvent of the second solution;
7. Dissolving at least one third polymer from the second remaining solid component in the selected solvent at the third higher temperature to form a third solution; and
8. Separating the at least one third polymer from the selected solvent of the third solution.

5.2.11 Compatibilization

Multi-material multilayers usually include polymer blends that are mechanically unstable (38).

For this reason, the compatibilization technique uses chemical substances to increase the mechanical stability of the multilayers and further recycle all layers in a single stream without separation (39).

These methods make polymer blends less immiscible by adding another component to the structures (5).

5.2.12 Feedstock Recycling

An alternative emerging treatment for difficult to recycle plastic waste might be producing chemicals and fuels such as monomer or hydrocarbon feedstocks.

Processes known to perform this are gasification, pyrolysis, fluid-catalyzed cracking, and hydrocracking (5). Pyrolysis may become an active technology for plastic waste, such as MMPP, which is complex to depolymerize (40). The process occurs at high temperatures and in the absence of oxygen when polymers are broken down and, with further processing, results in petrochemical feedstock such as naphtha or diesel (5). This process, known as chemical recycling, includes thermolysis through catalysts to selectively generate gases, liquids, or waxes, which are developments that go beyond mechanical recycling (41).

At the moment, this is not a widespread recycling practice due to energy costs (42). However, the technology might evolve as a future feasible option for plastics or plastics-containing waste that is difficult to be materially recycled from an economical and systematic perspective, opening new routes and opportunities.

A study developed by Solis and Silveira (42) assessed eight technologies regarding feedstock recycling for mixed plastic waste and their technology readiness level (TRL).

The most developed techniques are pyrolysis, catalytic cracking, and conventional gasification, whereas other emerging technologies noticeably demand more investments to be at a commercial stage (42).

5.2.13 *Closed-Loop Recycling*

Poly(ethylene terephthalate) (PET) is a favored candidate for closed-loop recycling for new packaging applications (14). A PET-recycling process was investigated with respect to its cleaning efficiency and the compliance of the PET recyclate with food law.

The key technology of the investigated PET-recycling process to remove contaminants consists of a predecontamination-extruder combination. At the end of the recycling process, there is either a pelletizing system or downstream equipment to produce preforms or flat sheets.

Therefore, the process has two process options, an inline production of PET preforms and a batch option producing PET pellets. In the case of possible misuse of PET bottles by the consumer, the inline process produces higher concentrations in the bottle wall of the recyclate containing preforms (14).

Owing to the dilution of the PET output material by large amounts of uncontaminated PET, the batch option is the less critical process in terms of consumer protection. Regarding an appropriate testing procedure for the evaluation of a bottle-to-bottle recycling process, both process options have their own specific requirements with respect to the design of a challenge test (14).

A closed-loop control system for recycling products, such as food service ware, has been presented (43). It uses the following steps:

1. Organizing a number of entities in a group that together use sizable quantities of disposable PS food service ware;
2. Collecting the used service ware from the group, while avoiding mixing the used service ware with other service ware PS products that did not emanate from the PS service ware of the group;
3. Establishing a facility for washing, grinding and pelletizing the used PS food service ware; and
4. Transferring the pelletized used PS to an entity that combines the pelletized used PS with virgin PS in proportions allowed under government regulations and arranges for the conversion of the mixture to recycled PS food service ware, and for the sale of the service ware to the entities in the group, and continuing the cycle over again.

Figure 5.2 illustrates a closed-loop continuous system for recycling food service ware, such as expanded PS dishes and trays used in institutions such as public schools, colleges, universities, industrial catering, etc.

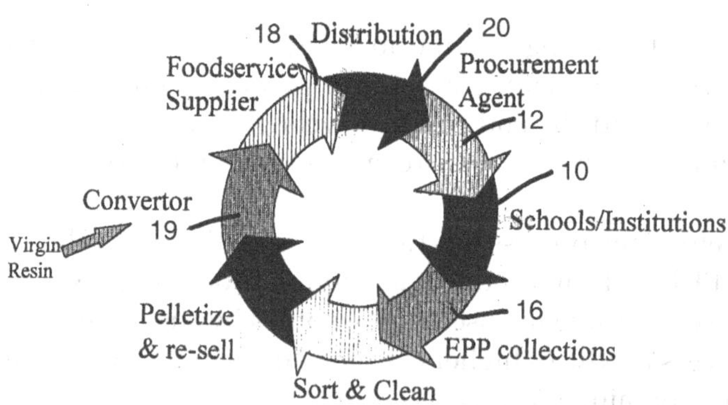

Figure 5.2 Closed-loop continuous system for recycling food service ware (43).

In this illustration, a number of participants are identified, such as schools and other institutions that employ disposable food service ware 10; a system coordinator or procurement agent 12 may oversee the collection and conversion of the waste products, and the converter 19 utilizes the washed and pelletized product and mixes it with additional designated virgin raw material to manufacture the new food service ware. Various functions are performed by each of the participants and some of their respective tasks may be carried out by one or more entities. From the converter 19, newly manufactured food service ware is ultimately supplied back to the schools or institutions 10 where the cycle is repeated over and over again. During each cycle of the system, additional virgin material is combined with the converted used service ware. In one embodiment, the newly manufactured food service ware is transported from the converter 19 to a distributor 20, and then to the procurement agent 12, to meet the supply demands of the school or institution 10. In some embodiments, the recycled material may be tested to detect any extractives or microbial contamination. This testing may occur once the finished product is produced and/or may also occur after the used food service ware is cleaned (43).

5.2.14 *Supercritical Ethanol*

Different properties, such as barriers to gas, light, flavor and water vapor, as well as flexibility, are necessary for the production of food packages. These properties may be obtained by combining different polymers. In spite of the advantages achieved with the use of multilayer films, the recycling process of such material is a challenging task (22). An alternative for recycling of a quite common type of food packaging was presented. Here, PE, aluminum and PET were used as a multilayer film. The multilayer packages were delaminated with acetone.

The depolymerization of PET took place by the reaction with ethanol under supercritical conditions. The obtained products in this reaction are diethyl terephthalate and ethylene glycol. The diethyl terephthalate is easily obtained through precipitation in water. On the other hand, ethylene glycol could be purified from ethanol and water by distillation.

Diethyl terephthalate was obtained as the main product, presenting high purity and yield of 80%. Also, metallic aluminum was obtained. The optimal reaction time was 120 *min* (22).

5.3 Plastic Waste Generation

The packaging industry requires materials that are lightweight so as to reduce the amount used to package a product, thus reducing product weight for transportation cost savings as well as reducing the amount of end-of-life packaging waste material (44). Plastics have managed to fulfill this role very well and have remained unchallenged.

For example, a study revealed that the replacement of glass bottles with plastic bottles for beverage packaging in airlines resulted in savings of over USD 1 million in fuel costs as a result of the weight loss (45).

For recycling of plastics, the most important steps are collection and sorting. Sorting involves the selection of waste which is suitable for recycling using manual or automated means. This can be done at the source of the waste and by using separate collection systems or at facilities by mechanical processing and sorting of mixed waste (46).

5.4 Recycled Plastics in Food Contact

Sustainability in packaging is definitely needed. One way for consumer product companies to reach their sustainability goals is to have their packages made from recycled plastic. There is a wide variety of legislation covering recycled plastics for use in food contact applications. Some jurisdictions allow it, and some have denied it (47).

5.4.1 *Improving Safety and Quality*

The use of recycled plastics for food packaging will likely increase due to the high price of crude oil, a feedstock for the petrochemicals used in the manufacture of plastics, as well as the recent advances in plastic recycling technologies that allow improved quality and safety of post-consumer recycled plastics intended for food packaging applications.

Some efforts at improving the safety and quality of recycled plastics for food contact applications have been presented (48).

Several types of plastics are used in packaging. To provide a consistent national system that facilitates the collection and recycling of post-consumer plastics through the normal channels for collecting recyclable materials from household waste, the Society of the Plastics Industry (SPI) developed a resin identification code system for common plastics used in the manufacture of containers. The SPI code system applies to six homogeneous plastics and one heterogeneous material category.

To improve the recyclability of plastic packaging for food contact use, controls on the source of post-consumer plastics as well as limitations on the use of the specific recycled article are needed to ensure the quality and safety of the recycled materials for the proposed food contact application (48).

5.5 Enzyme Discovery and Engineering for Sustainable Plastic Recycling

The drastically increasing amount of plastic waste is causing an environmental crisis that requires innovative technologies for recycling post-consumer plastics to achieve waste valorization while meeting environmental quality goals (49).

Biocatalytic depolymerization mediated by enzymes has emerged as an efficient and sustainable alternative for plastic treatment and recycling.

A variety of plastic-degrading enzymes have been discovered from microbial sources. At the moment, protein engineering has been exploited to modify and optimize plastic-degrading enzymes.

The recent trends and up-to-date advances in plastic-degrading enzymes have been reviewed.

Also, future research prospects and challenges have been discussed (49).

Plastic-degrading enzymes that have been discovered by the metagenomic approach are shown in Table 5.6.

5.6 Special Compositions

Globally, there are only low recycling rates for single-use plastic packaging materials with only 14% of plastic packaging being col-

Table 5.6 Plastic-degrading enzymes (49).

Enzyme	Metagenomic source	
	Function-based screening	
Esterase	Seawater	(50)]
Esterase	Marine environments, soils, and waste treatment facilities	(51)
Esterase	Sphagnum bog Fosmid	(52)
Esterase	Compost Plasmid	(53)
Esterase	Compost Fosmid	(54)
Cutinase	Leaf-branch compost	(55)
	Sequence-based screening	
PHB depolymerase	Biofilms on marine plastics	(56)
PET hydrolase	Marine and terrestrial environments	(57)
Poly(urethane) esterase	Landfill	(58)

lected for recycling and only 5% of it successfully recycled into new plastic materials (59, 60).

Common single-use plastic packaging waste materials polluting the environment in order of quantity include drinking bottles, bottle caps, food wrappers, grocery bags, lids, straws, stirrers and foam take-out containers (61, 62).

The food industry is encouraged to adopt the reduce, reuse and recycle waste control options rather than adopt the easy direct disposal method. This waste that is disposed in landfills results in methane gas creation, a greenhouse gas that surpasses carbon dioxide by more than 2000% in causing global warming (63).

It has been observed that plastic plays a vital role in maintaining food safety, extending shelf life and minimizing food waste. However, it is currently not possible to recycle the multilayer plastic packaging widely used throughout the food industry, and there are currently no viable alternatives offering the same level of protection. Unless possible substitutes for multilayer plastics offering the same level of food protection can be developed, there will be detrimental effects on food quality, safety and shelf life, which will lead to increased food waste, additional food costs and a reduction in the variety and availability of certain foods.

The principal conventional polymers and bioplastics used in food packaging in terms of their resources or raw materials, properties, uses and applications have been described in a chapter of a monograph (64).

Food packaging plastics represent an important part of total plastic waste that ends up in landfills or accumulates in the environment when they are not properly managed. For counteracting this contamination problem, plastic recycling is the best alternative. One of the challenges to guaranteeing recycling efficiency is the adequate polymer separation of plastic waste streams because of contaminants or the presence of polymers different than the target recycled plastic, which can cause processability problems or variation of its original properties (64).

5.7 Bottle Recycling

The disposal of plastic waste has become an important issue due to the potential environmental impact of improper waste disposal. The packaging with PET accounted for 44.7% of single-serve beverage packaging in the US in 2021, and 12% of global solid waste (65).

In comparison, aluminum cans accounted for 39%, glass for 11%, and high-density poly(ethylene) (HDPE) for 3.4%.

5.7.1 PET Bottles

The PET bottle has several properties that make it an environmentally friendly choice. It has good barrier properties as its single-layer, mono-material composition allows it to be more easily recycled. Compared to glass, the PET bottle is lightweight and has a lower carbon footprint in production and transportation.

Poly(ethylene terephthalate) (PET) is one of the most widely used packaging materials for beverages. Due to its excellent transparency, light weight, gas and water barrier properties, impact strength, UV resistance, and stabilty, the production and use of PET bottles for beverage packaging has consistently increased worldwide. The use of PET is a recyclable solution with performance benefits that are not available in alternative packaging options such as glass bottles, aluminum cans, paperboard cartons, and other plastics.

Bottles made from PET account for 67% of the market share between water, carbonated soft drinks, energy drinks, tea, and coffee.

Since PET is a petroleum-based polymer, it does not readily decompose when released into Earth's environment via plastic waste leakage. In 2015, the global plastic waste generated was approximately 141 million tons (66). Packaging waste buried in landfills can still contribute to air, water, and soil pollution. Additionally, plastic leakage into landfills consumes available landfill space. It must be noted that the percentage of plastic in the landfill by volume is higher than that by weight (67).

Incineration of plastic packaging waste avoids consumption of landfill space and generates energy, but with the drawbacks of emissions creation and air pollution.

The Great Pacific Garbage Patch is another example of unwanted plastic ending up in an undesirable place. Pollution from plastic waste, therefore, has been acknowledged as a major global environmental issue.

Approximately 27 *Mt* (million tons) of plastic (18.5% of US plastic waste) from wraps, bottles, and jars were discarded into landfills. Only 4.5% of plastic packaging was recycled (68).

According to European Economic Area data for the EU in 2019, 34.4 *kg* of plastic packaging waste was generated per EU inhabitant, on average.

The EU recycled at a rate of 41% (14.1 *kg*) per inhabitant, on average (69). Globally, PET accounted for 12% of total solid waste.

The EU Commission for the Environment has conceived pathways by which member countries may significantly reduce plastic waste leakage. These pathways include concepts for incentivizing change in consumer behavior, improvement of waste management involving waste collection, waste separation and recycling, as well as the restriction of wastes accepted into landfills (70).

Source reduction, also known as waste prevention, is defined by the EPA as a change in the design, manufacturing, purchasing or use of materials or products (including packaging) to reduce their amount or toxicity before they become a municipal solid waste (65).

To control and reduce the plastic waste, source reduction can be implemented utilizing principles of waste management practice, i.e., reuse, reduce, redesign, and recycle packaging in combination with packaging innovation.

Bottle reuse through returnable bottle systems has been widely practiced in South America and in several European countries.

However, in most regions, PET bottles are intended for single-use packaging and are disposed of after first use. To reduce the environmental impact from packaging waste and to drive sustainability in plastic packaging, municipal recycling programs have been implemented to capture recyclable material streams (65).

5.7.2 *Compatibilization of PET and PLA*

At the moment, products from partly or fully recycled materials are preferably produced. In addition, biopolymers are used besides or instead of petroleum-based polymers (71). Similarly to petroleum-based plastics, most biopolymers are used by the packaging industry (72).

However, due to their function, they have a very short lifetime of a few weeks. Therefore, they become waste in a short time (66).

Poly(lactic acid) (PLA) is one of the most popular biodegradable biopolymers used in the packaging industry to produce films, sheets, bottles and foams (72–76).

The separation of mixed PET and PLA bottles in the post-consumer plastic waste stream is difficult and expensive with conventional methods. Manual sorting by visual appearance cannot be done because in most cases both PET and PLA bottles are transparent, therefore they look very similar. Their density is also very similar, i.e., 1.2–1.3 $g\,cm^{-3}$ for PLA and 1.3–1.4 $g\,cm^{-3}$ for PET, and higher than that of water. Therefore, a traditional water-based float-sink separation process is not effective (77, 78).

It was demonstrated that even small amounts of PLA have a significant negative effect on the properties of PET (79, 80).

At the processing temperature of PET, PLA already degrades, which leads to a yellowing of the product. Moreover, PET and PLA are thermodynamically immiscible; therefore, holes, peaks or clusters can appear in the products.

In addition, the glass transition temperature of the two polymers is also different, resulting in opaqueness or haziness in PLA-contaminated PET products (81).

5.8 Recycling of Post-Consumer Polyolefins

5.8.1 *Predictive Models*

The use of post-consumer polyolefins is increasing due to the pledges of multiple organizations to use higher concentrations of recycled materials in packaging (82). However, direct food contact applications are challenging, with various regulations around the world mandating minimum safety requirements, which are often determined by the condition of use (microwave, cold storage) and food type (fatty, aqueous, dry).

Various predictive models that have been developed to study contaminant migration from packaging to food contact articles have been described.

Also, multiple techniques and methods were examined that have been employed to quantify and assess the safety of recycled materials (e.g., United States and Europe), which serve as a base for the regulatory framework for many other countries. Also, the presence of different contaminants during polymer processing is discussed due to potential problems from the presence of harmful compounds affecting human health or degradation of material properties (82).

5.8.2 *Safety Concerns*

Plastic recycling is one of the important ways to mitigate plastic pollution. However, chemicals present in recycled plastics are one of the key qualities affecting their potential uses and deserve more attention.

The application of 475 migrants coming from 15 post-consumer recycled polyolefins was confirmed by direct-immersion solid-phase microextraction coupled with gas chromatography mass spectrometry, and atmospheric pressure gas chromatography coupled with quadrupole time-of-flight mass spectrometry. About 60% of them might not be a risk to humans because they were food additives/components or saturated hydrocarbons, fatty acyls, or prenol lipids. Most of them had a molecular weight between 150 *Da* and 210 *Da*, though high concern substances with high molecular weight (e.g., octocrylene) implied that high molecular weight surrogates are required to study the efficiency of recycling processes for polyolefins. The mean predicted octanol/water partition coefficient

(XLogP) was about 6.5 and 3.5 for 95% ethanol and 3% acetic acid food simulants, respectively.

Octocrylene, 1-tetradecene, 1-dodecene, dodecyl acrylate, 2,4-di-*tert*-butylphenol, 1,4-benzenedicarboxylic acid, diethyl ester, benzenamine, 2,4-dichloro- and diethyl phthalate were of high concern depending on the potential food contact use of the materials.

The here presented results are informative and can be of great help to recyclers and lawmakers for recycling polyolefins for safe food contact use (83).

5.9 Recycling of Multi-material Multilayer Plastic Packaging

The recycling of multi-material multilayer plastic packaging (MMPP) is complex since the recycling industry is not able to identify, sort and separate the diverse layers with current standard technologies (20).

Therefore, MMPP is usually sorted as mixed plastic waste in post-consumer streams (also called other types of plastics) and frequently incinerated with energy recovery in European countries, which does not prioritize recycling as a circular economy practice (84).

In low-income countries, these packaging materials are usually disposed of at dumpsites or in landfills and a fraction also leaks into the environment due to the common lightweight and short lifespan (85).

5.9.1 Fast-Moving Consumer Goods

Multi-material multilayer plastic packaging (MMPP) is widely applied in fast-moving consumer goods combining functionalities of distinct materials (86). These packaging structures can enhance properties, such as resource-use efficiency and barrier performance, leading to consequential benefits like a prolonged shelf life. Nevertheless, they represent a challenge for existing recycling systems, confronting circular economy principles.

In a study, the future of recycling technologies for MMPP was assessed. Future scenarios were identified, including (86):

1. High-performance material recycling,
2. Recycling into hydrocarbons,
3. Business as usual, and
4. Downcycling.

In-depth interviews and a feedback survey were methods used to validate the scenario matrix while defining some expectations towards the future.

Circular economy models have been studied as solutions to current and future environmental challenges (87).

The analysis showed that distinct technologies will develop unevenly in different parts of the world (86).

Packaging for fresh food products may consist of four to seven layers of different components (88). A wide range of substances with different physical and chemical characteristics form multi-material multilayers (30).

Various polymers are applied, such as polyolefins PE and PP and chemical variants (HDPE, low-density poly(ethylene), linear low-density poly(ethylene)), or polyesters such as PET and PLA. Other materials are ethylene vinyl alcohol, poly(urethane), and poly(vinylidene chloride). These multilayer multi-material films and sheets are developed by co-extrusion or lamination techniques (89).

An overview of the main materials, their functionalities and applications are shown in Table 5.7.

5.9.1.1 HDPE Multilayer Composites

A method for recycling post-consumer multilayer wastes containing PET, PE and aluminum has been described (22).

Composite materials were prepared by using ground multilayer films as reinforcement phase in recycled HDPE matrix. The PET and aluminum worked as reinforcing fillers to establish better mechanical properties, while PE improved the adhesion between matrix and fillers. The composites were characterized by scanning electron microscopy, thermal analysis, water uptake, tensile, flexural and Izod impact tests.

The addition of a multilayer matter in the PE matrix provided a material with excellent mechanical properties such as tensile impact

Table 5.7 Materials applied for multi-material multilayer plastic packaging (86).

Material	Layer Functionality	Application Types
PE	- heat-sealable - moisture barrier - combination with gas/aroma barriers (e.g., PA, EVOH)	Beverage cartons (usually contains LDPE layer)
PP	- moisture barrier - coated with heat seal coatings (PVDC, acrylate) for mechanical strength -combined with gas/aroma barriers	Toiletries, cosmetics, microwavable food packaging
PET	-gas/aroma barrier -moisture barrier -mechanical strength -heat-sealable	Food packaging such as bottles, meat, or cheese
PS	-gas permeability -printability -combined with gas/aroma barriers	Printable outside layer for fresh food packaging
PVC	-gas/aroma barrier -mechanical strength	Stretch wrap
PA	-gas/aroma barrier -mechanical strength -heat-sealable	Thermoformed food packaging
EVOH	-oxygen barrier	Oxygen-sensitive food packaging

Compound	Abbreviation
Poly(ethylene)	PE
Poly(propylene)	PP
Poly(ethylene terephthalate)	PET
Poly(styrene)	PS
Poly(vinyl chloride)	PVC
Poly(amide)	PA
Ethylene vinyl alcohol	EVOH

Table 5.7 (cont) Materials applied for multi-material multilayer plastic packaging (86).

Material	Layer Functionality	Application Types
EVA	-adhesive -heat-sealable	Interlayer to bond PE to PVC
PUR	-adhesive (dry bond lamination) -heatsealable	Cosmetics, pharmaceutical packaging
PVDC	-gas/aroma barrier -moisture barrier -abrasion protection	Film wraps
PLA	-gas/aroma barrier -moisture barrier -mechanical strength	Cups, containers, take-out packaging
Bio-PE	-heat-sealable -resistant to hydrolysis/oxidation -flexibility	Similar uses as fossil fuel PE (uses the same recycling stream)
PHA	-gas/aroma barrier -moisture barrier -mechanical strength	Applications in development (may replace PET)
PEF	-gas/aroma barrier -moisture barrier -mechanical strength	Applications in development (may replace PET)

Compound	Abbreviation
Ethylene vinyl acetate	EVA
Poly(urethane)	PU
Poly(vinylidene chloride)	PVDC
Poly(lactic acid)	PLA
Bio-Poly(ethylene)	Bio-PE
Poly(hydroxyalkanoate)s	PHA
Poly(ethylene furanoate)	PEF

strength and high elasticity. The presence of different components in the multilayer packaging, which was the main obstacle preventing its recycling, became the key feature to develop a material with superior properties compared to those presented by commercial products. The composites presented a low water uptake and high thermal stability. One of the greatest advantages in the preparation of such composites was the possibility to range the formulation PE/filler up to 50% of filler, providing final materials with tailored mechanical properties according to their composition.

The approach presented offers new possibilities for process and product development from flexible multilayer wastes contributing to the mitigation of future environmental issues (22).

5.9.2 Recyclability Enhancement of Food Containers

Pizza boxes and take-out food containers are particular culprits when it comes to recycling (90). Proper recycling of these containers relies in part on proper curbside preparations. In the case of take-out food containers, proper washing with soap and hot water is required to remove grease. Pizza boxes cannot be washed, and generally have the bottom portion torn off and discarded separately.

To solve these problems, a special food container was fabricated. The food container includes a first interior face having disposed thereon a first protective layer for protecting the first interior face from exposure to contaminants that would interfere with recycling of the food container. The first protective layer is affixed to the first interior face by an adhesive interface so as to be peelable from the first interior face.

The protective layer offers numerous advantages, including protecting the surface of the container from contaminants, and in some cases increasing the shelf life of the content of the container (90).

5.10 Future Trends in Summary

In this chapter the frontiers of plastics recycling were discussed. Following the circular economy hierarchy of reduce and reuse, recycling is the third layer to close the loops for materials and decouple their value from consumption.

Polymeric materials, i.e., plastics, are in principle well suited for recycling, as they can be reprocessed with relatively low energy input as a material, cleaved back into their monomers or converted back to feedstock.

Today, these approaches still fall short in quantitatively diverting waste towards reuse. Here, the industry challenges were described for recycling back into high-value applications of polymers, and the portfolio of existing and emerging technology solutions.

Sustainable design of products and polymers, recycling technologies, appropriate business models, and enabling technologies converge at the frontier of plastics recycling, for a transformation of the industry towards circularity and greenhouse gas emission neutrality.

References

1. H. Mangold and B. von Vacano, The frontier of plastics recycling: Rethinking waste as a resource for high-value applications, *Macromolecular Chemistry and Physics* 2100488 (2022).
2. K. Bellmann and A. Khare, European response to issues in recycling car plastics, *Technovation* 721–734 (1999).
3. R.K. Scalice, D. Becker, and R.C. Silveira, Developing a new compatibility table for design for recycling, *Product: Management and Development* 141–148 (2009).
4. Z.O. Schyns and M.P. Shaver, Mechanical recycling of packaging plastics: A review, *Macromolecular Rapid Communications* 2000415 (2021).
5. K. Ragaert, L. Delva, and K. Van Geem, Mechanical and chemical recycling of solid plastic waste, *Waste Management* 24–58 (2017).
6. D. Feldman, Polyblend compatibilization, *Journal of Macromolecular Science, Part A* 587–605 (2005).
7. W. Beitz, G. Pahl, and K. Grote, Engineering Design: A Systematic Approach, *MRS Bulletin* (1996).
8. S. Wagner and M. Schlummer, Legacy additives in a circular economy of plastics: Current dilemma, policy analysis, and emerging countermeasures, *Resources, Conservation and Recycling* 104800 (2020).
9. A.N. Gaduan, K. Singkronart, C. Bell, E. Tierney, C. Burgstaller, and K.-Y. Lee, Mechanical upcycling immiscible polyethylene terephthalate-polypropylene blends with carbon fiber reinforcement, *ACS Applied Polymer Materials* 3294–3303 (2022).

10. B. Geueke, K. Groh, and J. Muncke, Food packaging in the circular economy: Overview of chemical safety aspects for commonly used materials, *Journal of Cleaner Production* (2018).
11. M. Farhoodi, Z. Emam-Djomeh, M.R. Ehsani, and A. Oromiehie, Migration of model contaminants (ethylene glycol, DEHA and DEHP) from PET bottles into Iranian yogurt drink, *e-Polymers* (2008).
12. C.P. Rader, S.D. Baldwin, D.D. Cornell, G.D. Sadler, and R.F. Stockel, *Plastics, Rubber, and Paper Recycling: A Pragmatic Approach*, ACS Publications, (1995).
13. M. Huber and R. Franz, Studies on contamination of post consumer plastics from controlled resources for recycling into food packaging applications, *Deutsche Lebensmittel-Rundschau (Germany)* 328 (1997).
14. R. Franz and F. Welle, Recycled poly(ethylene terephthalate) for direct food contact applications: Challenge test of an inline recycling process, *Food Additives & Contaminants* (2002).
15. F.L. Bayer, Polyethylene terephthalate recycling for food-contact applications: Testing, safety and technologies: A global perspective, *Food Additives & Contaminants* 111–134 (2002).
16. C. Nerin, J. Albinana, M.R. Philo, L. Castle, B. Raffael, and C. Simoneau, Evaluation of some screening methods for the analysis of contaminants in recycled polyethylene terephthalate flakes, *Food Additives & Contaminants* 668–677 (2003).
17. A.E. Goulas, P. Zygoura, and A. Karatapanis, Migration of di(2-ethylhexyl) adipate and acetyltributyl citrate plasticizers from food-grade PVC film into sweetened sesame paste (halawa tehineh): Kinetic and penetration study, *Food Chem. Toxicol.* (2007).
18. M. Farhoodi, S.M.A. Mousavi, R. Sotudeh Gharebagh, Z. Emam-Djomeh, and A. Oromiehie, Migration of silicon from nanocomposite packaging materials into acidic food simulant, in *Advanced Materials Research*, Vol. 622 873–877. Trans Tech Publ., (2013).
19. M. Farhoodi, S.M. Mousavi, R. Sotudeh-Gharebagh, Z. Emam-Djomeh, and A. Oromiehie, Migration of aluminum and silicon from PET/clay nanocomposite bottles into acidic food simulant, *Packaging Technology and Science* 161–168 (2014).
20. O. Horodytska, F. Valdés, and A. Fullana, Plastic flexible films waste management – A state of art review, *Waste Management* 413–425 (2018).
21. S. Zhang, K. Luo, L. Zhang, X. Mei, S. Cao, and B. Wang, Interfacial separation and characterization of Al–PE composites during delamination of post-consumer Tetra Pak materials, *Journal of Chemical Technology & Biotechnology* 1152–1159 (2015).
22. S.L. F?varo, A.R. Freitas, T.A. Ganzerli, A.G.B. Pereira, A.L. Cardozo, O. Baron, E.C. Muniz, E.M. Girotto, and E. Radovanovic, PET and aluminum recycling from multilayer food packaging using supercritical ethanol, *Journal of Supercritical Fluids* (2013).

23. M. Perick, Method for recycling plastics from a film packaging bag, EP Patent 3 059 061, assigned to Mondi AG, July 19, 2017.
24. F. Lovis, H. Seibt, and S. Kernbaum, Method and apparatus for recycling packaging material, US Patent 10 682 788, assigned to Saperatec GmbH, June 16, 2020.
25. G. Dodbiba and T. Fujita, Progress in separating plastic materials for recycling, *Physical Separation in Science and Engineering* (2004).
26. J. Shibata, S. Matsumoto, H. Yamamoto, E. Kusaka, and Pradip, Flotation separation of plastics using selective depressants, *International Journal of Mineral Processing* 127–134 (1996).
27. F. De Santis and R. Pantani, Optical properties of polypropylene upon recycling, *The Scientific World Journal* (2013).
28. I.A. Ignatyev, W. Thielemans, and B. Vander Beke, Recycling of polymers: A review, *ChemSusChem* 1579–1593 (2014).
29. G. Curtzwiler, K. Vorst, J.E. Danes, R. Auras, and J. Singh, Effect of recycled poly(ethylene terephthalate) content on properties of extruded poly(ethylene terephthalate) sheets, *Journal of Plastic Film & Sheeting* 65–86 (2011).
30. A. Mieth, E. Hoekstra, and C. Simoneau, Guidance for the identification of polymers in multilayer films used in food contact materials, *European Commission JRC Technical Reports* 1–70 (2016).
31. A.J. Hadi, G.F. Najmuldeen, and I. Ahmed, Polyolefins waste materials reconditioning using dissolution/reprecipitation method, *APCBEE Procedia* 281–286 (2012). 2nd International Conference on Chemistry and Chemical Process (ICCCP 2012) May 5-6, 2012.
32. T.W. Walker, N. Frelka, Z. Shen, A.K. Chew, J. Banick, S. Grey, M.S. Kim, J.A. Dumesic, R.C. Van Lehn, and G.W. Huber, Recycling of multilayer plastic packaging materials by solvent-targeted recovery and precipitation, *Science Advances* eaba7599 (2020).
33. Y.-B. Zhao, X.-D. Lv, and H.-G. Ni, Solvent-based separation and recycling of waste plastics: A review, *Chemosphere* 707–720 (2018).
34. M. Schlummer, T. Fell, A. Mäurer, and G. Altnau, The role of chemistry in plastics recycling: A comparison of physical and chemical plastics recycling, *Kunststoffe International* 34–37 (2020).
35. I. Georgiopoulou, G.D. Pappa, S.N. Vouyiouka, and K. Magoulas, Recycling of post-consumer multilayer Tetra Pak® packaging with the selective dissolution-precipitation process, *Resources, Conservation and Recycling* 105268 (2021).
36. D.S. Achilias, A. Giannoulis, and G.Z. Papageorgiou, Recycling of polymers from plastic packaging materials using the dissolution-reprecipitation technique, *Polym Bull* (2009).
37. E.B. Nauman and J.C. Lynch, Polymer recycling by selective dissolution, US Patent 5 278 282, assigned to Rensselaer Polytechnic Institute, January 11, 1994.

38. X.-M. Xie and X. Sui, Compatibilization of polymer blends by the addition of random copolymers in A.R. Ajitha and S. Thomas, (Eds.), *Compatibilization of Polymer Blends* chapter 5, 145–177. Elsevier, (2020).
39. F. Mostafapoor, A. Khosravi, A. Fereidoon, R. Khalili, S.H. Jafari, H. Vahabi, K. Formela, and M.R. Saeb, Interface analysis of compatibilized polymer blends in A.R. Ajitha and S. Thomas, (Eds.), *Compatibilization of Polymer Blends* chapter 12, 349–371. Elsevier, (2020).
40. T. Thiounn and R.C. Smith, Advances and approaches for chemical recycling of plastic waste, *Journal of Polymer Science* 1347–1364 (2020).
41. J.M. Garcia and M.L. Robertson, The future of plastics recycling, *Science* 870–872 (2017).
42. M. Solis and S. Silveira, Technologies for chemical recycling of household plastics – A technical review and TRL assessment, *Waste Management* 128–138 (2020).
43. M. Forrest and D. Schneider, Closed-loop control system for recycling products, US Patent 20 050 027 555, assigned to Michael W. Forrest, David Schneider, February 03, 2005.
44. L.K. Ncube, A.U. Ude, E.N. Ogunmuyiwa, R. Zulkifli, and I.N. Beas, An overview of plastic waste generation and management in food packaging industries, *Recycling* 12 (2021).
45. K.S. Marsh, Polymer and plastic packaging in *Reference Module in Food Science*. Elsevier, (2016).
46. C. Cimpan, A. Maul, M. Jansen, T. Pretz, and H. Wenzel, Central sorting and recovery of MSW recyclable materials: A review of technological state-of-the-art, cases, practice and implications for materials recycling, *Journal of Environmental Management* 181–199 (2015).
47. M.H. Filyaw, Recycled plastics in food contact in J.S. Baughan, (Ed.), *Global Legislation for Food Contact Materials*, Woodhead Publishing Series in Food Science, Technology and Nutrition chapter 8, 193–202. Elsevier, 2nd edition, (2022).
48. V. Komolprasert and A. Bailey, Recycled plastics for food applications: Improving safety and quality in E. Chiellini, (Ed.), *Environmentally Compatible Food Packaging*, Woodhead Publishing Series in Food Science, Technology and Nutrition chapter 13, 326–350. Woodhead Publishing, (2008).
49. B. Zhu, D. Wang, and N. Wei, Enzyme discovery and engineering for sustainable plastic recycling, *Trends in Biotechnology* 22–37 (2022).
50. A. Tchigvintsev, H. Tran, A. Popovic, F. Kovacic, G. Brown, R. Flick, M. Hajighasemi, O. Egorova, J.C. Somody, D. Tchigvintsev, et al., The environment shapes microbial enzymes: five cold-active and salt-resistant carboxylesterases from marine metagenomes, *Applied Microbiology and Biotechnology* 2165–2178 (2015).

51. A. Popovic, T. Hai, A. Tchigvintsev, M. Hajighasemi, B. Nocek, A.N. Khusnutdinova, G. Brown, J. Glinos, R. Flick, T. Skarina, et al., Activity screening of environmental metagenomic libraries reveals novel carboxylesterase families, *Scientific Reports* 44103 (2017).
52. C.A. Müller, V. Perz, C. Provasnek, F. Quartinello, G.M. Guebitz, and G. Berg, Discovery of polyesterases from moss-associated microorganisms, *Applied and Environmental Microbiology* e02641–16 (2017).
53. D. Mayumi, Y. Akutsu-Shigeno, H. Uchiyama, N. Nomura, and T. Nakajima-Kambe, *Applied Microbiology and Biotechnology* 743–750 (2008).
54. C.-H. Kang, K.-H. Oh, M.-H. Lee, T.-K. Oh, B.H. Kim, and J.-H. Yoon, A novel family VII esterase with industrial potential from compost metagenomic library, *Microbial Cell Factories* 1–8 (2011).
55. S. Sulaiman, S. Yamato, E. Kanaya, J.-J. Kim, Y. Koga, K. Takano, and S. Kanaya, Isolation of a novel cutinase homolog with polyethylene terephthalate-degrading activity from leaf-branch compost by using a metagenomic approach, *Applied and Environmental Microbiology* 1556–1562 (2012).
56. L.J. Pinnell and J.W. Turner, Shotgun metagenomics reveals the benthic microbial community response to plastic and bioplastic in a coastal marine environment, *Frontiers in Microbiology* 1252 (2019).
57. D. Danso, C. Schmeisser, J. Chow, W. Zimmermann, R. Wei, C. Leggewie, X. Li, T. Hazen, and W.R. Streit, New insights into the function and global distribution of polyethylene terephthalate (PET)-degrading bacteria and enzymes in marine and terrestrial metagenomes, *Applied and Environmental Microbiology* e02773–17 (2018).
58. I. Gaytán, A. Sánchez-Reyes, M. Burelo, M. Vargas-Suárez, I. Liachko, M. Press, S. Sullivan, M.J. Cruz-Gómez, and H. Loza-Tavera, Degradation of recalcitrant polyurethane and xenobiotic additives by a selected landfill microbial community and its biodegradative potential revealed by proximity ligation-based metagenomic analysis, *Frontiers in Microbiology* 2986 (2020).
59. P. Dauvergne, Why is the global governance of plastic failing the oceans?, *Global Environmental Change* 22–31 (2018).
60. J.N. Hahladakis and E. Iacovidou, Closing the loop on plastic packaging materials: What is quality and how does it affect their circularity?, *Science of the Total Environment* 1394–1400 (2018).
61. C. Giacovelli, *Single-Use Plastics: A Roadmap for Sustainability*, International Environmental Technology Centre, (2018).
62. T.Y. Chitaka, V. Russo, and H. von Blottnitz, In pursuit of environmentally friendly straws: A comparative life cycle assessment of five straw material options in South Africa, *The International Journal of Life Cycle Assessment* 1818–1832 (2020).

63. N. Singh, D. Hui, R. Singh, I.P.S. Ahuja, L. Feo, and F. Fraternali, Recycling of plastic solid waste: A state of art review and future applications, *Composites Part B: Engineering* 409–422 (2017).
64. E. Velásquez, M. Guerrero Correa, L. Garrido, A. Guarda, M.J. Galotto, and C. López de Dicastillo, Food packaging plastics: Identification and recycling in *Recent Developments in Plastic Recycling* 311–343. Springer, (2021).
65. P. Benyathiar, P. Kumar, G. Carpenter, J. Brace, and D.K. Mishra, Polyethylene terephthalate (PET) bottle-to-bottle recycling for the beverage industry: A review, *Polymers* 2366 (2022).
66. R. Geyer, J.R. Jambeck, and K.L. Law, Production, use, and fate of all plastics ever made, *Science Advances* e1700782 (2017).
67. S.E. Selke, Recycling of plastic packaging in the United States, *Packaging Technology and Science* 93–98 (1988).
68. Anonymous, Facts and figures about materials, waste and recycling: Plastics: Material-specific data, Technical report, EPA, Washington, D.C., (2018).
69. Anonymous, Recycled 41EPA, Washington, D.C., (2018).
70. E. Commission, (Ed.). *Communication from the Commission to the Council, the European Parliament, the European Economic and Social Committee and the Committee of the Regions: Civil Society Dialogue Between the EU and Candidate Countries*, Vol. 290, Brussels, Belgium, (2005). European Commission, European Commission.
71. D. Gere and T. Czigany, Future trends of plastic bottle recycling: Compatibilization of PET and PLA, *Polymer Testing* 106160 (2020).
72. F. Gironi and V. Piemonte, European bioplastics. bioplastics market data 2018, *Energy sources, Part A: Recovery, Utilization and Environmental Effects* 1949–1959 (2011).
73. W. Chow, E.L. Teoh, and J. Karger-Kocsis, Flame retarded poly(lactic acid): A review, *Express Polymer Letters* 396–417 (2018).
74. Á. Kmetty, K. Litauszki, and D. Réti, Characterization of different chemical blowing agents and their applicability to produce poly (lactic acid) foams by extrusion, *Applied Sciences* 1960 (2018).
75. H. Chen, X. Yu, W. Zhou, S. Peng, and X. Zhao, Highly toughened polylactide (PLA) by reactive blending with novel polycaprolactone-based polyurethane (PCLU) blends, *Polymer Testing* 275–280 (2018).
76. M.M. Mazidi, R. Berahman, and A. Edalat, Phase morphology, fracture toughness and failure mechanisms in super-toughened PLA/PB-g-SAN/PMMA ternary blends: A quantitative analysis of crack resistance, *Polymer Testing* 380–391 (2018).
77. D. Gere and T. Czigany, Recycling of mixed poly (ethylene-terephthalate) and poly (lactic acid), in *MATEC Web of Conferences*, Vol. 253 02005. EDP Sciences, (2019).

78. M.R. Gent, M. Menendez, J. Toraño, and I. Diego, Recycling of plastic waste by density separation: Prospects for optimization, *Waste Management & Research* 175–187 (2009).
79. F.P. La Mantia, L. Botta, M. Morreale, and R. Scaffaro, Effect of small amounts of poly (lactic acid) on the recycling of poly (ethylene terephthalate) bottles, *Polymer Degradation and Stability* 21–24 (2012).
80. D.D. Cornell, Biopolymers in the existing postconsumer plastics recycling stream, *Journal of Polymers and the Environment* 295–299 (2007).
81. L. Alaerts, M. Augustinus, and K. Van Acker, Impact of bio-based plastics on current recycling of plastics, *Sustainability* 1487 (2018).
82. V.S. Cecon, P.F. Da Silva, G.W. Curtzwiler, and K.L. Vorst, The challenges in recycling post-consumer polyolefins for food contact applications: A review, *Resources, Conservation and Recycling* 105422 (2021).
83. Q.-Z. Su, P. Vera, C. Ner?n, Q.-B. Lin, and H.-N. Zhong, Safety concerns of recycling postconsumer polyolefins for food contact uses: Regarding (semi-)volatile migrants untargetedly screened, *Resources, Conservation and Recycling* 105365 (2021).
84. K. Kaiser, M. Schmid, and M. Schlummer, Recycling of polymer-based multilayer packaging: A review, *Recycling* 1 (2017).
85. J.N. Hahladakis and E. Iacovidou, An overview of the challenges and trade-offs in closing the loop of post-consumer plastic waste (PCPW): Focus on recycling, *Journal of Hazardous Materials* 120887 (2019).
86. C.T. de Mello Soares, M. Ek, E. Ostmark, M. Gallstedt, and S. Karlsson, Recycling of multi-material multilayer plastic packaging: Current trends and future scenarios, *Resources, Conservation and Recycling* 105905 (2022).
87. P. Ghisellini, C. Cialani, and S. Ulgiati, A review on circular economy: The expected transition to a balanced interplay of environmental and economic systems, *Journal of Cleaner Production* 11–32 (2016).
88. T.I. Butler and B.A. Morris, PE-based multilayer film structures in J.R. Wagner, (Ed.), *Multilayer Flexible Packaging*, Plastics Design Library chapter 17, 281–310. William Andrew Publishing, 2nd edition, (2016).
89. J.R. Wagner, (Ed.), *Multilayer Flexible Packaging*, Plastics Design Library, William Andrew, 2nd edition, (2016).
90. K. Suzuki, S. Suzuki, S. Suzuki, and W.D. Bishop, Jr., Recyclability enhancement of food containers, US Patent Application 20 130 118 946, assigned to Kiyan Suzuki, Shabnam Suzuki, Sana Suzuki, William David Bishop, Jr., May 16, 2013.

6

Renewable Energy

6.1 Plastic Waste

6.1.1 Strategy to Sort and Recycle Plastic Waste

An optimal strategy to sort and recycle plastic waste as a renewable energy resource with maximizing economic feasibility and mitigating environmental pollution has been suggested (1).

To derive the optimal sorting and recycling strategies of plastic waste, a novel optimization model was developed. The model calculates the overall profit by subtracting the profit of recycling plastic from the total annualized cost. Then the model is used to identify the optimal strategy to sort and recycle plastic waste as a renewable energy resource in mixed-integer nonlinear programming that maximizes the overall profit.

In the here derived optimal sorting and recycling strategy, HDPE is recycled to produce downgrade plastic; high-density poly(ethylene) (HDPE), poly(propylene) (PP), and poly(styrene) (PS) are recycled as pyrolysis oil; and poly(ethylene terephthalate) (PET) is recycled to produce refuse plastic fuel.

The derived optimal case can significantly increase the overall profit by about 3,137% (i.e., 35 US$ per 1 *kg* of recycled plastic), and 492% (i.e., 29 US$ per 1 *kg* of recycled plastic) compared to the conventional case in South Korea and Japan respectively (1).

Economically viable sorting and recycling of plastic will yield a cheap and abundant source of valuable chemicals and renewable energy (2).

To solve these challenges, this work optimizes the sorting and recycling of plastic waste as a renewable energy resource, maximizing the overall profit of a system that sorts and recycles plastic waste, by considering economic feasibility and environmental preservation. The novelty of this work can be summarized as follows (1):

1. It is the first attempt to optimize the plastic waste sorting and recycling system by deriving an optimal sorting and recycling strategy of plastic waste as a renewable energy resource maximizing economic feasibility and mitigating environmental pollution caused by landfilled plastic and air pollutants according to incineration of plastic.
2. The results allow increasing the plastic waste recycling by maximizing the overall profit of the plastic waste sorting and recycling system with mitigation of environmental pollution. Thus, it can also yield tremendous economic improvements and environmental protection effects.
3. Finally, this study provides valuable insight to increase the cleanliness and cost-effectiveness of production, and help to protect the environment.

In summary, an optimal strategy was found to sort and recycle plastic waste as a renewable energy resource for maximizing economic feasibility and mitigation of environmental pollution caused by landfilled plastic and air pollutants according to incineration of plastic.

In the so-derived optimal sorting and recycling strategy, high-density poly(ethylene) is recycled to produce downgrade plastic (1). Low-density poly(ethylene), PP and PS are recycled as pyrolysis oil. PET is recycled to produce refused plastic fuel. The derived optimal case can increase overall profit by 3,137% compared to the conventional case in South Korea, and 492% compared to the conventional case in Japan, and also decrease the emission of air pollutants. Thus, a valuable insight was provided into the many recycling industries of waste plastics to achieve clean production, cost-effectiveness, and environmental protection. Strategies, such as operating condition optimization, were proposed to reduce the cost of each sorting method. Thus, in a future study, it is crucial to consider the detailed proposed strategies in deriving the optimal solution to maximize economic feasibility (1).

6.1.2 *Gasification System for Hydrogen Production*

Synthetic plastics are basically petroleum-based synthetic organic polymers with properties that make them suitable for a wide range of applications, including packaging, construction, home and sporting goods, automobiles, electronics, and agriculture.

Every year, more plastic-based products are manufactured, with the majority of that amount going toward single-use items such as shopping bags, cups, and straws (3).

So, a huge amount of plastic waste has been accumulated over the years without an effective disposal method. For effective disposal of plastic waste, certain infrastructure is required, including incinerator facilities, recycling capacity, circular economy infrastructure, and waste management and disposal systems. As the infrastructure required for efficient and effective disposal of plastic wastes in an environmentally benign manner is not accessible for many countries, leakage of plastic waste into rivers and oceans happens. Irresponsible disposal of plastic waste can have a detrimental effect on our ecosystem and biodiversity (4).

Fox and Stacey (3) exhaustively examined thermal decomposition through gasification and pyrolysis as the most environmentally friendly plastic waste management techniques. The study concluded that both technologies are economically viable and would create a huge environmental impact if implemented on a large scale. Although pyrolysis could be marginally more environmentally friendly as the process produces less carbon emissions, gasification is more economically viable. Furthermore, the study suggested gasification could be even more economically viable if the excess energy generated in the production process was utilized to generate useful outputs, which is what is being proposed in this study.

Multigeneration systems are systems designed to integrate multiple subsystems to produce different useful outputs. While energy used to run multigeneration systems coming from different types of sources, including fossil fuels like natural gas, a sustainable multigeneration system is normally designed to run on renewable energy resources such as solar, geothermal, wind or hydro (5). In this study, a sustainable waste-to-energy multigeneration system was developed that will couple sustainable energy resources with intricate systems to effectively utilize plastic waste.

This study will focus on producing hydrogen as a clean fuel since hydrogen-driven energy systems are a viable solution for concurrent renewable energy as they may be used in lieu of fossil-based fuels. Many potential applications for hydrogen exist in the effort to decarbonize processes and industries, as greenhouse gases are not released during hydrogen combustion. Furthermore, hydrogen has one of the highest energy contents in comparison to other fuels. Currently, the bulk of hydrogen requirement is for chemical processes, including the use of hydrogen as a processing agent in some chemical processes.

However, it is expected that hydrogen is used as a sustainable alternative fuel in other applications, including transportation vehicles.

The production and distribution costs of hydrogen energy must be decreased if its use is to become practical and sustainable.

Recent advancements in hydrogen production and transportation suggest that the cost of environmentally friendly hydrogen fuel will be lower than that of traditional fuels in the near future.

A multigeneration system to fully exploit plastic waste will be created in this project (6, 7). The waste-to-energy system will be created using the 3S model (source, system, and service), utilizing plastic wastes and geothermal energy as the energy sources. Using only renewable energy resources as the main source of energy for the multigeneration system adds sustainability to the system. Coupling renewable energy sources with waste-to-energy systems produces the maximum possible environmental impact as we are extracting energy from waste material using a renewable and benign energy source. In the proposed system, the waste source is plastic waste, which is not only wasted energy but also an environmental risk that is causing damage to every aspect of the environment. The primary goal of this article is to design and evaluate a multigeneration waste-to-energy system that relies on plastic waste gasification.

Plastic wastes are a valuable source of energy as they are essentially composed of concentrated hydrocarbons. As such, utilizing plastic waste as an energy source is exploited in many research studies. Mohan investigated the use of waste plastic oil to replace fossil fuels, particularly in diesel engines (8). The plastic wastes were pyrolyzed to produce synthetic oil, and a four-stroke diesel engine was fed with varying concentrations of waste plastics, oil alternatives, and diesel blends. The engine performance was further evaluated under various load conditions, speeds, and compression ratios. Their results showed that up to 40% alternative diesel mixtures can operate reliably, but their thermal efficiency was lower than diesel and their emissions were higher.

Devasahayam and Strezov studied the effects of using biomass and plastic wastes in the process of thermal decomposition of magnesium carbonate (9).

Plastic waste has the potential to produce high-energy synthetic gas because it is an energy-dense material.

The co-gasification of plastic wastes with biomass wastes was studied (10) as an attractive energy recovery option. The simulations were based on the analysis of the co-gasification process for different combinations of plastic wastes and biomass wastes at different temperatures. Their results indicate that increasing the proportion of plastic wastes in the gasification feedstock increases the gas heating value and hence its potential to be used as a more efficient fuel.

A dual-stage gasification for plastic wastes to produce syngas was investigated (11). Using 900 *g* of nickel-loaded activated carbon, the highest hydrogen content in the resultant syngas was 30%.

A multigeneration system was presented for producing hydrogen and community requirements in a specific area of Pakistan (12), utilizing energy from the sun and municipal solid waste. The municipal solid wastes included minimal amounts of plastic wastes, which were turned into syngas using a biomass boiler. The study reported energy efficiency of 88.6%.

The effect of increasing the gasification reactor temperature on the production of syngas was investigated (13). The study was based on modeling the gasification of waste poly(ethylene) (PE) and PP plastic. The study showed that as the temperature of the gasification reactor rises, so does the energy content of the produced syngas. By boosting the gasification temperature to a maximum limit of 900°C, the graph struck a plateau. Hence, raising the gasification reactor temperature more would not be beneficial.

He *et al.* studied the gasification of PE from municipal solid wastes (14). The gasification reactor was coupled with a NiO catalyst for enhanced vapor yield. The experiment was maintained between 700 and 900°C. Steam was introduced to the gasifier at a rate of 1.33

times the PE feed rate. The results showed hydrogen output of 36.98% and CO output of 27.35%.

Geothermal energy could be utilized to drive the multigeneration system to produce multiple useful outputs, including hydrogen as an environmentally benign fuel (15). The proposed integrated system integrates geothermal energy and plastic waste gasification to produce syngas that is processed through steam-iron reforming for hydrogen production.

The steam-iron reforming process for hydrogen production was studied (16). The effects of adding two catalysts to the iron regenerator were shown. Catalysts were copper and chromium. The catalyst lowers the iron-steam reforming running temperature from 1000 *K* to 570 *K*. The study also found that while iron-steam reforming significantly improves hydrogen purity compared to methane reforming, using the catalyst has no effect on hydrogen purity and maintains its high effectiveness. The experimental results also showed that repeating the process has no noticeable effect on efficiency.

6.1.2.1 Multigeneration Systems

Despite the fact that multigeneration systems are quite efficient and thus a well-researched subject in the literature, only few people consider incorporating a waste-to-energy system into multigeneration (17). Furthermore, despite being a major environmental concern, plastic waste is rarely studied when it comes to multigenerational waste-to-energy. Some studies have classified plastic waste as municipal solid waste, but this does not address the issue directly. Furthermore, multigenerational systems typically produce output tailored to either industrial or residential needs. As the five outputs of the proposed system are hydrogen, fresh water, heating, domestic hot water, and electric power, it meets both communal and industrial needs. The key novelties of the proposed system can be summarized as follows (17):

1. The present integrated system is newly designed in a way to treat plastic wastes and utilize geothermal energy sources for multigenerational purposes.
2. The plastic wastes are converted into useful energy forms along with hydrogen.

3. The useful outputs of the present system may potentially serve to cover industrial and residential needs.
4. The present system decomposes wastes using gasification, rather than traditional thermal decomposition methods, i.e., incineration, to minimize the environmental impact.
5. The present system is evaluated through energy and exergy efficiencies by conducting some parametric studies to identify what operating conditions provide better results.

6.1.3 *Renewable Energy and Plastic Waste Recycling*

Circular economy is an approach that focuses on economic and environmental sustainability dimensions that arise in response to increasing natural resource consumption, especially of nonrenewable resources, as a result of world population increases (18).

To achieve a circular economy, industrial activities can be designed for reducing resource usage and minimizing waste generation through process improvements, increased recycling of products and materials used, and transitioning to the use of renewable energy sources

Shifting to more efficient, circular business models can also provide a competitive advantage by improving the performance and reducing the costs (18).

6.1.4 *Greenhouse Gas Emission*

The chemical industry is one of the most important industries in terms of greenhouse gas emissions (19). It has the highest energy demand of all industrial sectors as it uses fossil fuels, i.e., coal, oil, and gas, for energy as well as feedstock for the production of platform petrochemicals.

The most central processes in petrochemical value chains (including production of olefins and aromatics through steam cracking as well as ammonia and methanol synthesis) use fossil raw materials and are extremely energy intensive. The production of the platform chemicals used for downstream production of more complex products, such as fertilizers and plastics, require very high operational temperatures and pressures to manufacture. Emissions are generated on production sites through fossil fuel combustion as well as in many of the fundamental processes for producing key chemicals, but also in the external production of heat and power used by the industry.

Developments in the global chemical industry have rapidly reduced the greenhouse gas emissions through the use of renewable energy (19).

6.2 Future Trends in Summary

An optimal strategy to sort and recycle plastic waste as a renewable energy resource with maximizing economic feasibility and mitigating environmental pollution has been suggested.

To derive the optimal sorting and recycling strategies of plastic waste, a novel optimization model was developed. The model calculates the overall profit by subtracting the profit of recycling plastic from the total annualized cost. Then the model is used to identify the optimal strategy to sort and recycle plastic waste as a renewable energy resource in mixed-integer nonlinear programming that maximizes the overall profit.

Circular economy is an approach that focuses on economic and environmental sustainability dimensions that arise in response to increasing natural resource consumption, especially of nonrenewable resources, as a result of world population increases.

To achieve a circular economy, industrial activities can be designed for reducing resource usage and minimizing waste generation through process improvements, increased recycling of products and materials used, and transitioning to the use of renewable energy sources.

Shifting to more efficient, circular business models can also provide a competitive advantage by improving the performance and reducing the costs.

References

1. J. Lim, Y. Ahn, and J. Kim, Optimal sorting and recycling of plastic waste as a renewable energy resource considering economic feasibility and environmental pollution, *Process Safety and Environmental Protection* 685–696 (2023).

2. G. Gadaleta, S. De Gisi, S.M.C. Binetti, and M. Notarnicola, Outlining a comprehensive techno-economic approach to evaluate the performance of an advanced sorting plant for plastic waste recovery, *Process Safety and Environmental Protection* 248–261 (2020).
3. J.A. Fox and N.T. Stacey, Process targeting: An energy based comparison of waste plastic processing technologies, *Energy* 273–283 (2019).
4. O. Oner and I. Dincer, A unique solar and biomass-based system for integrated production of electricity, heat, freshwater, hydrogen and ethanol, *Energy Conversion and Management* 116115 (2022).
5. I. Dincer and C. Zamfirescu, Multigeneration systems in I. Dincer and C. Zamfirescu, (Eds.), *Advanced Power Generation Systems* chapter 9, 517–573. Elsevier, Boston, (2014).
6. M.M. Ismail and I. Dincer, Development of an integrated waste to energy multigeneration system based on plastic wastes for sustainable cities, *Sustainable Cities and Society* 104079 (2022).
7. I. Dincer and A. Abu-Rayash, *Energy Sustainability*, Academic Press, (2019).
8. R.K. Mohan, J. Sarojini, U. Rajak, T.N. Verma, and Ü. Ağbulut, Alternative fuel production from waste plastics and their usability in light duty diesel engine: Combustion, energy, and environmental analysis, *Energy* 126140 (2023).
9. S. Devasahayam and V. Strezov, Thermal decomposition of magnesium carbonate with biomass and plastic wastes for simultaneous production of hydrogen and carbon avoidance, *Journal of Cleaner Production* 1089–1095 (2018).
10. D. Buentello-Montoya, C. Duarte-Ruiz, and J. Maldonado-Escalante, Co-gasification of waste PET, PP and biomass for energy recovery: A thermodynamic model to assess the produced syngas quality, *Energy* 126510 (2023).
11. M.-H. Cho, Y.-K. Choi, and J.-S. Kim, Air gasification of PVC (polyvinyl chloride)-containing plastic waste in a two-stage gasifier using Ca-based additives and Ni-loaded activated carbon for the production of clean and hydrogen-rich producer gas, *Energy* 586–593 (2015).
12. T.A.H. Ratlamwala, S. Waseem, Y. Salman, and A.A. Bham, Geothermal and solar energy–based multigeneration system for a district, *International Journal of Energy Research* 5230–5251 (2019).
13. D. Saebea, P. Ruengrit, A. Arpornwichanop, and Y. Patcharavorachot, Gasification of plastic waste for synthesis gas production, *Energy Reports* 202–207 (2020).
14. M. He, B. Xiao, Z. Hu, S. Liu, X. Guo, and S. Luo, Syngas production from catalytic gasification of waste polyethylene: Influence of temperature on gas yield and composition, *International journal of Hydrogen Energy* 1342–1348 (2009).

15. M. Mahmoud, M. Ramadan, S. Naher, K. Pullen, M.A. Abdelkareem, and A.-G. Olabi, A review of geothermal energy-driven hydrogen production systems, *Thermal Science and Engineering Progress* 100854 (2021).
16. S. Takenaka, N. Hanaizumi, K. Otsuka, et al., Production of pure hydrogen from methane mediated by the redox of Ni-and Cr-added iron oxides, *Journal of Catalysis* 405–416 (2004).
17. M.M. Ismail and I. Dincer, A new renewable energy based integrated gasification system for hydrogen production from plastic wastes, *Energy* 126869 (2023).
18. W. Niyommaneerat, K. Suwanteep, and O. Chavalparit, Sustainability indicators to achieve a circular economy: A case study of renewable energy and plastic waste recycling corporate social responsibility (CSR) projects in Thailand, *Journal of Cleaner Production* 136203 (2023).
19. F. Bauer, J.P. Tilsted, S. Pfister, C. Oberschelp, and V. Kulionis, Mapping GHG emissions and prospects for renewable energy in the chemical industry, *Current Opinion in Chemical Engineering* 100881 (2023).

7

Methods of Characterization

7.1 Polymer Identification Techniques

The principal polymer identification techniques have been detailed that are applied to post-consumer plastic streams in order to improve the sorting and separation of plastics before recycling. Plastic waste recycling methods associated with single and multilayer polymer systems and their stages are described. Also, a comparative analysis of the properties variation of polymers used in food packaging with mechanical recycling is presented based on recent research studies, which shows the principal degradation reactions taking place for each polymer (1).

In 2018, the recycling rate of plastic packaging was only 40% (2). However, this is an overestimation since in practice this 40% accounts for the plastic waste which is collected for recycling. The actual net recycling rate of plastic packaging was in fact lower. A study on the post-consumer plastic packaging recycling network in the Netherlands, for instance, has indicated an effective recycling yield of only 26% instead of 50%, as was reported by the official authorities (3).

Besides improving recycling rates, the European Commission also stated that it is important for recyclers to work closely together with the chemical industry to help identify wider and high-value applications for their output. This together with the idea of a more circular economy implies the preference of the EU for closed-loop recycling scenarios, in which, for instance, food packaging is recycled to become a food contact material once more (2).

The recycling of food packaging materials can be estimated to be only 5% in a closed-loop process within the EU (4). This recycling process is mainly done by means of a mechanical recycling.

Here, washed and sorted plastic waste is remelted and processed into new food packaging. However, the recycling process may cause several issues such as the safe use of the recycled plastics as food packaging due to an increase in possible contamination sources, which might migrate into the food, thus imposing a risk to human health (2).

7.2 Identification of the Materials

7.2.1 *Spectroscopic Methods*

In the polymer recycling process, at least two steps might require an automatic identification tool:

1. The sorting of the plastics, and
2. The qualification of the regenerated product.

The efficiencies of reflectance FTMIR and FT-Raman were tested with respect to these identification applications. Although it was possible with both techniques to recognize most usual polymers, severe limitations of FTMIR were evidenced, such as a high sensitivity to the surface state and a weak (noisy) reflectance signal for CH bonds. On the other hand, the FT-Raman technique proved to be a rapid (1 *s* per object) and highly selective method, giving information even on the mineral fillers present in plastics (5).

7.2.2 *Real-Time Mass Spectrometry*

Distinguishing poly(amide) (PA) polymers is useful in directing monomer residual testing, product testing, and reverse engineering, but is analytically challenging. The applicability of direct analysis in real time mass spectrometry DART-MS for screening PA polymers was tested (6).

A DART ion source coupled to a quadrupole Orbitrap high-resolution mass spectrometer was employed for this study. Ten types of PA polymers and four retail samples were evaluated. The so obtained data for these samples and their repeatability was assessed across days/calibrations, operators, and equipment.

Ions corresponding to the cyclic or linear monomers and oligomers of each PA polymer were detected in each mass spectrum. Although similar mass spectra were obtained for PA6, PA66, and PA6/PA66 (polymer blends of PA6 and PA66), their direct analysis in real time tandem mass spectra were completely different.

The analysis was repeatable, and nearly identical DART tandem mass spectra were obtained on different days, by different operators, and with different equipment.

So, this technique could be successfully applied to commercially available samples (6).

7.2.3 *Optical Identification*

A technology for automatic sorting of plastics, based upon optical identification of fluorescence signatures of dyes incorporated in such materials in trace concentrations prior to product manufacturing, has been described (7).

Three commercial tracers were selected primarily on the basis of their good absorbency in the 310 *nm* to 370 *nm* spectral band and their identifiable narrow-band fluorescence signatures in the visible band of the spectrum when present in binary combinations. This absorption band was selected because of the availability of strong emission lines in this band from a commercial mercury (Hg) arc lamp and high fluorescence quantum yields of the tracers at this excitation wavelength band.

The plastics chosen for tracing and identification are high-density poly(ethylene), low-density poly(ethylene), poly(propylene), ethylene vinyl acetate, poly(vinyl chloride) and poly(ethylene terephthalate), and the tracers were compatible and chemically nonreactive with the host matrices and did not affect the transparency of the plastics. The design of a monochromatic and collimated excitation source by a sensor system is described and its performances in identifying and sorting plastics doped with tracers at a few parts per million concentration levels were evaluated. In an industrial sorting system, the sensor was able to sort 300 *mm* long plastic bottles at a conveyor belt speed of 3.5 m, s^{-1} with a sorting purity of around

95%. The limitation was imposed due to mechanical singulation irregularities at high speed and the limited processing speed of the computer used (7).

7.3 Future Trends in Summary

This chapter dealt with the methods of characterization of polymers. Also, the principal polymer identification techniques have been detailed that are applied to post-consumer plastic streams in order to improve the sorting and separation of plastics before recycling.

Plastic waste recycling methods associated with single and multilayer polymer systems and their stages were described. Also, a comparative analysis of the properties variation of polymers used in food packaging with mechanical recycling was presented based on recent research studies, which show the principal degradation reactions taking place for each polymer.

References

1. E. Velásquez, M. Guerrero Correa, L. Garrido, A. Guarda, M.J. Galotto, and C. López de Dicastillo, Food packaging plastics: Identification and recycling in *Recent Developments in Plastic Recycling* 311–343. Springer, (2021).
2. E. De Tandt, C. Demuytere, E. Van Asbroeck, H. Moerman, N. Mys, G. Vyncke, L. Delva, A. Vermeulen, P. Ragaert, S. De Meester, and K. Ragaert, A recycler's perspective on the implications of REACH and food contact material (FCM) regulations for the mechanical recycling of FCM plastics, *Waste Management* 315–329 (2021).
3. M. Brouwer, C. Picuno, E.U.T. van Velzen, K. Kuchta, S. De Meester, and K. Ragaert, The impact of collection portfolio expansion on key performance indicators of the Dutch recycling system for post-consumer plastic packaging waste, a comparison between 2014 and 2017, *Waste Management* 112–121 (2019).
4. L. Lugal, A. Grant, M. Cordle, and D. Fletcher, PET market in Europe: State of play, *Eunomia Reports* (2020).
5. J. Florestan, A. Lachambre, and N. Mermilliod, Recycling of plastics: Automatic identification of polymers by spectroscopic methods, *Resources, Conservation and Recycling* (1994).

6. Y. Abe, L.K. Ackerman, and M. Mutsuga, Rapid identification of polyamides using direct analysis in real time mass spectrometry, *Rapid Commun Mass Spectrom* (2020).
7. S.R. Ahmad, A new technology for automatic identification and sorting of plastics for recycling, *Environmental Technology* (2004).

8

Medical Uses

Plastics in medical devices have been detailed in a book (1). Here, properties, requirements, and applications are presented along with a comprehensive overview on the main types of plastics used in medical device applications.

The applications and properties that are most important in medical device design, such as chemical resistance, sterilization capability and biocompatibility, are the focus of the book. The roles of additives, stabilizers and fillers as well as the synthesis and production of polymers are covered and backed up with a wealth of data tables. The book also covers other key aspects in detail, including regulations, compliance, purchasing controls and supplier controls, and process validation (1).

8.1 Optical Applications

There are several monographs dealing with polymers that can be used for contact lenses and other optical applications (2–4).

The market for contact lenses is constantly growing, with more than 125 million consumers as of 2004 (5). The estimated global market size was 7.1 billion US dollars in 2015 (6). A search of the patent literature showed that more than 100 patents have been filed since 2000.

The applications of contact lenses range from corrective vision and therapeutics to cosmetic appearance (4, 7).

Within these applications comes the demands from the end user of the lenses, including length of wear, comfort, durability, ease of use, stability of vision, etc.

This also means that within the applications of contact lenses comes the demands from manufacturers, such as material costs, ease of production, and reliability of the contact lenses, etc. The demands from manufacturers determine the parameters of the material which scientists must focus their research on for developing contact lens materials. This premise has guided materials scientists, from the creation of glass scleral lenses in the 1930s to rigid, non-gas-permeable poly(methyl methacrylate) (PMMA) in the 1940s.

The 1960s and the 1970s ushered in hydrogel (i.e., polymer and silicone) lenses, with silicone hydrogel proving to be the most dominant kind of contact lens material today.

Commonly, the labels of *hard* or *soft* are used as blanket definitions of contact lenses (4, 7).

Hard contact lenses are rigid (durable), gas-permeable lenses, whereas soft contact lenses are made of flexible, high-water-content material. Hard lenses are often interchangeably referred to as rigid gas-permeable lenses. However, this is not strictly true. The first PMMA lenses could be classified as hard, whereas modern gas-permeable lenses are, in fact, more flexible, due to the incorporation of low-modulus components. So, they are more rigid, rather than hard. Another defining characteristic is that PMMA hard lenses have no oxygen permeability, whereas rigid gas-permeable lenses are permeable. On the other hand, a soft contact lens is a highly flexible, oxygen-permeable material often with high water content.

This flexibility means that soft contact lenses fit the shape of a user's eye much faster than a rigid lens. Soft contact lenses can be disposed of daily, weekly, or monthly. These blanket definitions of contact lenses can hint at their material properties, though not for certain. There is often an overlap in materials used between hard and soft lenses, such as silicone hydrogels and rigid gas-permeables. Although both use silicone materials, the difference lies in factors such as the gel network and water content. Derivatives therein can further diversify the range of possible contact lenses and their properties. The requirements for contact lenses are quite extensive, and there are a huge number of existing contact lenses on the market to reflect this; daily-disposable lenses, weekly/monthly lenses, special fitting lenses, and even lenses that can be worn overnight. Users demands can be described by several general parameters, such as comfort, wear time, handling (cleaning, ease of use), cost, and vision specifications (4, 7).

The advantages and the disadvantages of contact lens material classes are shown in Table 8.1.

Table 8.1 Advantages and disadvantages of contact lenses (8).

Material	Advantages	Disadvantages
PMMA	Inexpensive, well-understood polymer	No oxygen permeability, inflexible on the eye
RGP	High oxygen permeability, durable	Expensive regents, requires hydrophilic comonomer, can be abrasive
HEMA hydrogel	Inexpensive, biocompatible, abundant copolymer possibilities	Low oxygen permeability, protein deposition issues
Silicone hydrogel	High oxygen permeability, durable, comfortable	Expensive regents, requires hydrophilic comonomer, can be abrasive
PVA	Inexpensive, straightforward manufacturing, biocompatible	Low oxygen permeability, fixed water content

Short	Material long
PMMA	Poly(methyl methacrylate)
RGP	Rigid gas-permeable
HEMA hydrogel	2-Hydroxyethyl methacrylate hydrogel
Silicone hydrogel	Silicone hydrogel
PVA	Poly(vinyl alcohol)

8.2 Materials

Wearable electronics have received widespread attention in recent years, including electronic contact lenses with electrical components that provide the lenses with an added functionality. Many applications for electronic contact lenses have been proposed, such as lenses having glucose sensors for diabetic patients (9), and lenses containing an electroactive element having a dynamic aperture (10).

Electronic lenses show a potential application for the correction of vision errors, such as myopia control and presbyopia, where a continuous range of focus, i.e., from near distance to far distance, is desired.

8.2.1 *Hydrogel Contact Lenses*

Commercially available contact lenses made from hydrogels are preferred over lenses made from non-hydrogel materials because they are generally more comfortable (11). Hydrogel contact lenses are typically made by a cast molding process in which a polymerizable composition is dispensed into a contact lens mold and subjected to curing conditions, typically UV light or heat, that cause the monomer mixture to polymerize. The resulting lens is removed from the mold and hydrated to form a hydrogel, which typically consists of from about 20% to 60% water by weight. During the hydration process the lens swells appreciably in size. A non-swelling material, such as electronic components, incorporated into the lens during the curing step can cause an uneven swelling of the hydrogel material upon hydration, resulting in damaged or distorted lenses that are unsuitable for ophthalmic use.

8.2.1.1 *Silicone Hydrogels*

A silicone elastomer-silicone hydrogel hybrid contact lens has been developed that is made from a silicone elastomer layer comprising an anterior side and a posterior side and a silicone hydrogel layer having a percent swell of about -5% up to about 20% adhered to the posterior side of the silicone elastomer layer (11). A delamination-resistant bond is present between the silicone elastomer layer and the silicone hydrogel layer.

A method of manufacturing a silicone elastomer-silicone hydrogel hybrid contact lens has also been presented (11).

This is done by cast molding a first curable composition in a first mold assembly to form a first layer of the silicone elastomer-silicone hydrogel hybrid contact lens. The first mold assembly consists of a first mold member defining an anterior side of the first layer and a second mold member defining the posterior side of the first layer. The first mold assembly is disassembled to provide the first layer adhered to only one of the first and second mold members.

A second curable composition is cast molded in a second mold assembly to form a second layer of the silicone elastomer-silicone hydrogel hybrid contact lens. The second mold assembly consists of the mold member to which the first layer is adhered and a third mold member. The second mold assembly is disassembled to provide a silicone elastomer-silicone hydrogel hybrid contact lens consisting of (11):

1. A silicone elastomer layer comprising an anterior side and a posterior side,
2. A silicone hydrogel layer adhered to the posterior side of the silicone elastomer layer, and
3. A delamination-resistant bond between the silicone elastomer layer and the silicone hydrogel layer.

Silicone hydrogels exhibiting desired combinations of physical and mechanical properties have been developed (12). These hydrogels are formed from a reactive monomer mixture including at least one *N*-alkyl methacrylamide, and at least one silicone-containing component. These silicone hydrogels may also contain hydrophilic components, crosslinking agents and toughening monomers. These silicone hydrogels are useful in preparing biomedical devices, ophthalmic lenses, and contact lenses.

The silicone hydrogels may contain toughening agents. Toughening agents are monomers whose corresponding homopolymers exhibit glass transition temperatures higher than 40°C. When added to the reactive mixture they improve the elongation of the resulting silicone hydrogels. Examples of such monomers are shown in Table 8.2.

Table 8.2 Toughening agents (12).

Compound	Compound
Methyl methacrylate	*tert*-Butyl methacrylate
Isobornyl methacrylate	Cyclohexyl methacrylate
Styrene	Substituted styrenes
N-4-Vinylbenzyl-*N*-alkyl acetamides	*N*-4-Vinylbenzyl pyrrolidone

The reaction mixture may contain additional reactive or non-reactive components such as UV absorbers, visible-light absorbers, photochromic compounds, pharmaceuticals, nutriceuticals, antimicrobial substances, tints, pigments, copolymerizable and non-polymerizable dyes, and release agents (12).

Usually the reactive components are mixed in a diluent to form a reaction mixture. Suitable diluents are shown in Table 8.3.

Table 8.3 Diluents (12).

Compound	Compound	Compound
Ethanol	1-Butanol	1-Decanol
1-Hexanol	1-Octanol	2-Butanol
2-Hexanol	2-Octanol	1-Pentanol
1-Propanol	2-Pentanol	2-Propanol
1-Dodecanol	Isopropanol	*tert*-Butanol
Decanoic acid	Octanoic acid	Dodecanoic acid
2-Ethyl-1-butanol	*tert*-Amyl alcohol	*tert*-Butoxyethanol
1-Ethoxy-2-propanol	2-Methyl-2-pentanol	2-Octyl-1-dodecanol
3-Methyl-3-pentanol	3,3-Dimethyl-2-butanol	
3,7-Dimethyl-3-octanol		
Disopropylamino ethanol		
1-*tert*-Butoxy-2-propanol		
2-(Diisopropylamino)ethanol		

A polymerization initiator is preferably included in the reaction mixture.

Initiators include compounds that generate free radicals at moderately elevated temperatures, and photoinitiators, such as aromatic α-hydroxy ketones, alkoxyoxybenzoins, acetophenones, acylphosphine oxides, bisacylphosphine oxides, and diketones with tertiary amines, or mixtures thereof. Initiators and photoinitiators are shown in Table 8.4 and in Figure 8.1.

The most preferred photoinitiator is bis(2,4,6-trimethylbenzoyl)-phenyl phosphine oxide (Irgacure 819®).

The preferred method for producing contact lenses is done by the molding of the silicone hydrogels, which is economical and enables precise control over the final shape of the hydrated lens.

For this method, the reaction mixture is placed in a mold having the shape of the final desired silicone hydrogel contact lens,

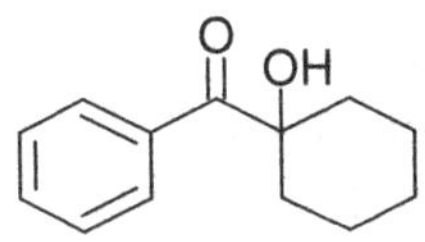

1-Hydroxycyclohexyl phenyl ketone

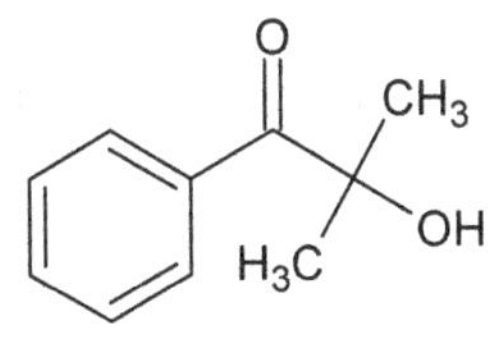

2-Hydroxy-2-methyl-1-phenyl-propan-1-one

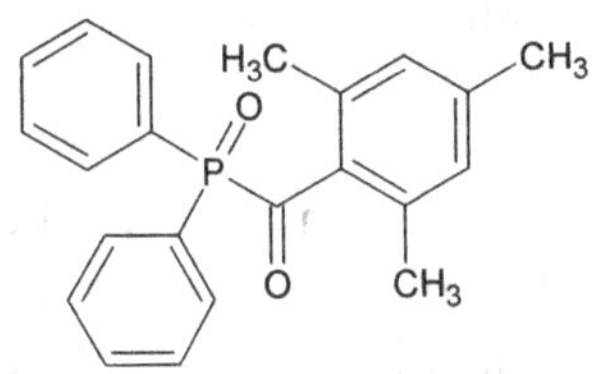

Bis(2,6-dimethoxybenzoyl)-2,4-4-trimethyl-pentyl phosphine oxide

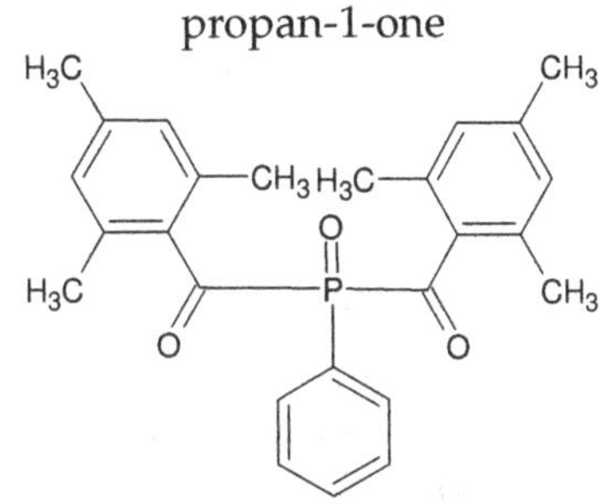

Bis(2,4,6-trimethylbenzoyl)-phenyl phosphine oxide

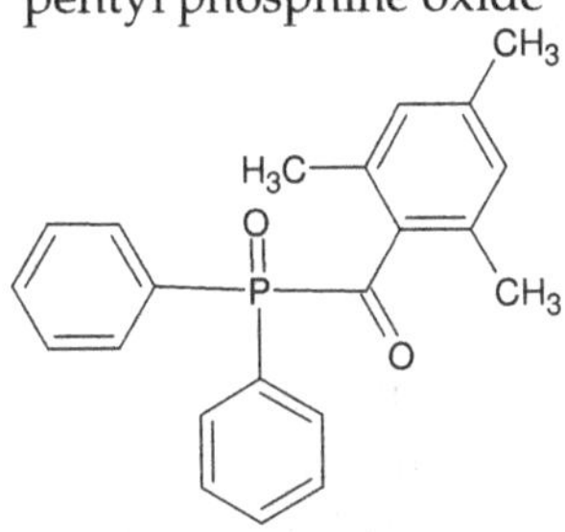

2,4,6-Trimethylbenzyldiphenyl phosphine oxide

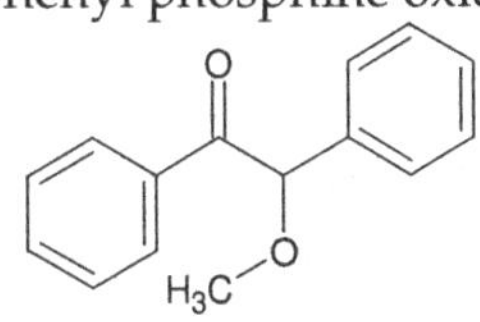

Benzoin methyl ester

Figure 8.1 Photoinitiators.

Table 8.4 Initiators and Photoinitiators (12).

Initiators
Lauryl peroxide, Benzoyl peroxide, Isopropyl percarbonate Azobisisobutyronitrile
Photoinitiators
1-Hydroxycyclohexyl phenyl ketone 2-Hydroxy-2-methyl-1-phenyl-propan-1-one Bis(2,6-dimethoxybenzoyl)-2,4-4-trimethylpentyl phosphine oxide Bis(2,4,6-trimethylbenzoyl)-phenyl phosphine oxide 2,4,6-Trimethylbenzyldiphenyl phosphine oxide 2,4,6-Trimethylbenzoyl diphenylphosphine oxide Benzoin methyl ester Camphorquinone and ethyl 4-(*N*,*N*-Dimethylamino)benzoate

and the reaction mixture is subjected to conditions whereby the monomers polymerize. Subsequently, the molded lens is treated with solvents to remove the diluent and ultimately replaced it with water or packaging solution, thereby producing a hydrated silicone hydrogel contact lens.

Some diluents can be removed by heat, vacuum or evaporation. In these cases, the lenses may be mechanically demolded without solvent. This method can be used to form contact lenses and has been described (13–16).

Ophthalmic lenses have a balance of properties which makes them particularly useful. These properties may be physical, mechanical, or biological in nature. Non-limiting physical properties of a contact lens include haze, water content, oxygen permeability, and contact angle. Non-limiting mechanical properties of a contact lens include Young's modulus, tensile strength, elongation at break, and toughness (12).

8.3 Surgery

Plastic surgery is a field of medicine that focuses on improving the appearance and function of various body parts through surgical procedures.

8.3.1 *ChatGPT*

With the advancements in technology, the use of artificial intelligence (AI) and natural language processing has become more prevalent in various fields, including plastic surgery. ChatGPT (Chat Generative Pre-Trained Transformer), is a large language model (LLM) developed by OpenAI. It has several potential uses in academic plastic surgery (17, 18).

One potential use of ChatGPT in plastic surgery research is in the analysis of patient data. With the ability to analyze vast amounts of patient data, ChatGPT can help identify patterns and trends that could lead to new insights in the field of plastic surgery. ChatGPT can also help predict patient outcomes and identify areas where surgical procedures can be improved.

Another potential use of ChatGPT is in the development of surgical techniques. ChatGPT can analyze data from surgical procedures and identify areas where the procedure can be improved or made more efficient. That can lead to the development of new techniques that could improve patient outcomes and reduce the risk of complications. Furthermore, it can help improve patient communication. With its natural language processing capabilities, ChatGPT can help research participants to understand the risk and benefits of participating in a study. It can also answer questions and provide information to patients before and after their surgery.

However, there are also several potential drawbacks of using ChatGPT in plastic surgery research (17). One of the main disadvantages is the lack of human interaction. Research participants may feel more comfortable and confident when speaking with a human rather than with a machine. Moreover, it has a significant potential for bias. While ChatGPT is designed to be objective, it is only as unbiased as the data it is trained on. If the data used to train ChatGPT are biased, it could lead to biased results and potentially harmful surgical procedures. Lastly, the cost of implementing ChatGPT can be a major barrier. The cost of training, maintaining, and implementing ChatGPT can be high and may not be feasible for all plastic surgery researchers.

In summary, ChatGPT has several potential uses in academic plastic surgery. As the technology continues to improve, it will be important to weigh the potential benefits and drawbacks of using

ChatGPT in plastic surgery research. Ultimately, a combination of human interaction and AI technologies may provide the best academic and patient care in plastic surgery.

Some caveats to be aware of when using LLMs such as ChatGPT in plastic surgery research are:

1. LLMs can help both researchers and journal editors write and edit articles, allowing them to focus on the research methodology and generating hypotheses rather than spending hours on writing and formatting. It can also help authors from different countries, where English is not their primary language, to write high quality manuscripts, and hence improve author diversity. However, if LLMs are to be used as a writing aid, it is imperative to review and edit all AI-generated texts in order to personalize and ensure its accuracy.
2. Using LLMs in research should be mentioned by all authors in their acknowledgement section in order to ensure transparency and honesty within the academic world.
3. LLMs are as accurate and unbiased as the data they are trained on, and therefore it is important for researchers to use them for literature searches with caution, and to cross-reference and critically analyze their generated information.
4. Journals should start specifying whether they accept LLM-generated text and in what capacity.

Finally, it is important to remember that as of now, the use of LLMs and their ethical considerations remain very elusive and unclear, specifically with how closely their work resembles that of humans. Therefore, academic plastic surgeons have the responsibility to familiarize themselves with this new technology and work together to develop guidelines for its use.

8.3.1.1 The Role of ChatGPT in Plastic Surgery

One of the primary reasons that ChatGPT presents an interesting ethical dilemma is that the chatbot itself is unable to fulfill authorship criteria and serve as an author (19). This presents challenges to using its content in professional publications.

ChatGPT is impressively able to execute tasks in seconds, reducing the time and resources needed to develop abstracts, manuscripts, and ideas. The overall burden of conceptualizing research projects may be significantly more manageable. However, authors should be cautious, and avoid the temptation to submit articles without appropriate references (19).

The limitations that ChatGPT presents because of its inherent inaccuracies have been detailed (19).

Along with possible inaccuracies and authorship issues, ChatGPT presents a unique challenge to journals and the research community. Anti-artificial intelligence programs may be a way to overcome this, because these programs are able to detect plagiarism and may play a heightened role for journals in the future.

8.3.2 3D Bioprinting

Three-dimensional (3D) bioprinting has gained widespread attention because it enables fine customization of the implants in the patient's surgical area preoperatively while avoiding some of the adverse reactions and complications of traditional surgical approaches (20).

The recent research advances in the application of 3D bioprinting in plastic surgery were reviewed. The printing process and the basic principles of 3D bioprinting technology were detailed. Also, the advantages and disadvantages of different bioprinting technologies were explained.

The main current 3D bioprinting technologies along with their principles and also their main advantages and disadvantages are explained in Table 8.5.

Also, the currently available bioprinting materials have been described. In addition, viable clinical applications and effects of 3D bioprinting in plastic surgery are detailed (20).

8.4 Polymer Implants

8.4.1 Dexamethasone

Based on the inflammatory etiology of radicular pain, analgesics and epidural steroid injections are a conventional intervention with

Table 8.5 Advantages and disadvantages of mainstream bioprinting technology (20).

Inkjet-based Bioprinting (21–24)	
Advantages	Disadvantages
Fast speed	Easy clogging of nozzles
Low cost	Limited ink viscosity
Strong simulation	Uneven ink size
	Poor sequence lines
Laser-based Bioprinting (25–29)	
Advantages	Disadvantages
High resolution	Photocrosslinker toxicity
High cell viability	Photo-induced gene mutation
Wide range of cell density	Expensive
Wide range of biomaterial viscosity	Time consuming
Extrusion-based Bioprinting (22, 30–35)	
Advantages	Disadvantages
Extensibility	Low resolution
Easy to operate	Low cell viability
High versatility	
Low price	
Wide range of cell density	
Wide range of biomaterial viscosity	

a therapeutic value in approximately half of the patients. These measures may alleviate pain enough to allow more physical activity.

Epidural steroid injections involve injecting a steroid, either alone or together with an analgesic, local anesthetic (lidocaine or bupivacaine), and/or saline directly into the problem area in the lower back to treat the inflammation that is irritating the nerve root. Dexamethasone, triamcinolone acetate, and methylprednisolone acetate are commonly used steroids. These compounds are shown in Figure 8.2.

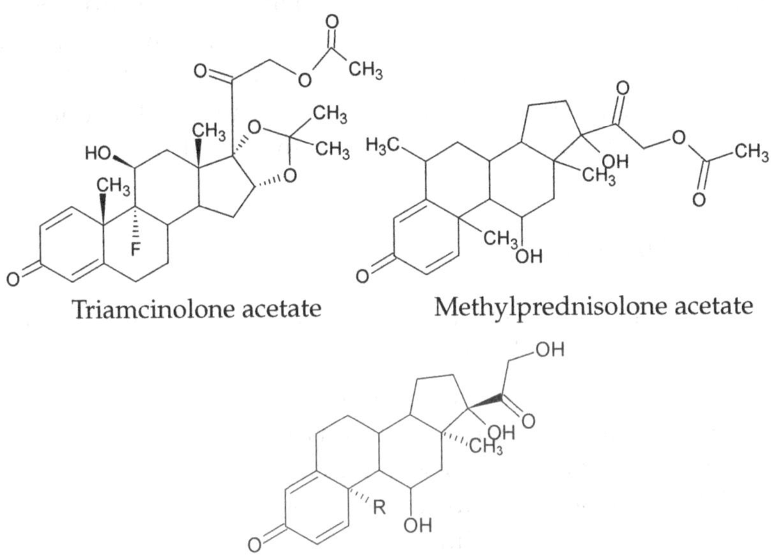

Triamcinolone acetate

Methylprednisolone acetate

Dexamethasone

Figure 8.2 Steroids.

A biodegradable drug implant for reducing, preventing or treating pain and/or inflammation in a patient has been provided.

The biodegradable drug implant contains dexamethasone and a poly(lactic acid) (PLA) poly(glycolic acid) (PGA) copolymer matrix, wherein the dexamethasone contains from about 50% to about 80% by weight of the drug implant.

Dexamethasone is a corticosteroid, also known as (8S, 9R, 10S, 11S, 13S, 14S, 16R, 17R)-9-Fluoro-11,17-dihydroxy-17-(2-hydroxyacetyl)-10,13,16-trimethyl-6,7,8,11,12,14,15,16-octahydro-cyclopenta[a]phenanthren-3-one, c.f. Figure 8.2.

8.5 Orthopedic Applications

Every year, many people suffer from bone fractures because of accidents or diseases (36). The majority of these fractures are too complicated to be treated with conventional medicine and must be mended surgically using non-degradable metal inserts. Such a treatment may result in refusal and protection against stress, but it may also require another surgical procedure to get rid of the metal inserts. Additionally, biodegradable metals can readily erode inside the human body and so come with certain complications.

8.5.1 Implant Metals

Some implant metals applications, their advantages, and disadvantages are shown in Table 8.6.

These aspects have directed scientists to find alternatives to metals. Some researchers began to focus on the field of polymers, which have shown significant promise in replacing metals. In orthopedics degradable polymeric fixation appliances have been studied to substitute metallic implants, eliminating stress protection and avoiding another implant removal surgery.

8.5.2 Bioabsorbable and Degradable Polymeric Implants

A new generation of bioabsorbable and degradable polymeric implants are now free from toxic and mutagenic effects. Nevertheless, these implants have several issues, including mechanical stiffness and strength limitations, unfavorable tissue responses, foreign body reactions, the late response of degraded tissue, and infection due to crystallinity and hydrophobicity.

It has been reported that alternative synthetic polymer implant materials are available that can be employed to avoid these problems.

Table 8.6 Implant metal applications (37).

Material	Advantages	Disadvantages	Usages
316l stainless steel	Excellent fabrication property, toughness, easily available and low cost, and accepted biocompatibility	High modulus, poor wear and corrosion resistance, stress shielding effect and allergic reaction	Bone plates, screws, pins, and wires, etc.
Co-Cr alloys	Better in terms of resistance to corrosion, fatigue and wear. High strength, long-term biocompatibility stress shielding effect, high modulus, toxicity due to Cr, Co and Ni ions release	Quite expensive, difficult to machine	Shorter term implants-bone plates and wires, total hip replacements
Ti alloys	Outstanding resistance to corrosion, lower modulus, stronger than stainless steels, light-weight, biocompatible	Poor wear resistance, poor bending ductility, expensive	Fracture fixation plates, fasteners, nails, screws, rods and wires, total joint replacement (TJR)
Mg alloys	Biocompatible, biodegradable, bioresorbable, lower density, Young's modulus of bone ($E = 10 - 30GPa$), low stress shielding effect, lightweight	Hydrogen evolution during degradation	Bone screws, plates, pins, stents, etc.

8.5.3 *Bone*

Bone is a natural compound in living tissue that contains roughly 30% matrix, 60% minerals, and 10% water (37).

Bone is one of the most critical human body components and has a supportive and protective function.

Bone has numerous functions, including providing a strong framework for protection of most vital tissues, and the binding site for tendons, ligaments and muscles [3]. Every year, many people suffer from bone fractures because of accidents or diseases. The majority of these fractures are too complicated to be treated with conventional medicine and must be mended surgically using non-degradable metal implants. Such treatment can cause refusal and protection against stress, but may also require another surgical procedure to get rid of the metal inserts. Additionally, this type of cure can worsen physical suffering and imposes a heavy financial burden on the patient.

8.6 Sutures

The use of biodegradable materials for medical applications has been studied extensively over the past four decades (38). The most common synthetic biodegradable polymers used in medical applications are poly(α-hydroxy acid)s, including PGA, PLA, and poly(dioxanone).

The minimal requirements of biomaterials intended for medical applications are nontoxicity, effectiveness, sterilizability, and biocompatibility. The oldest and broadest medical application of biodegradable polymers are as sutures (38). A suture is a rigid joint between hard elements of an organism, with or without significant overlap of the elements, found in a wide range of animals.

8.7 Biomedical Uses

8.7.1 *Coatings from Polysaccharides*

Polysaccharides exhibit important properties for biomedical targeted applications due to their stimuli responsiveness, mechanical toughness, low permeability, electrical conductivity, catalytic activity, biocompatibility, nontoxicity, bacteriostatic and antimicrobial properties. Coatings obtained from polysaccharides are one step closer to resembling the extracellular matrix and offer compositional individuality like inducing a particular cellular response. For instance, it is well known that hyaluronic acid interacts with a variety of receptors, including cell surface receptors (39).

Chitosan can be used as an antimicrobial polyelectrolyte in combination with synthetic polyelectrolyte to obtain antimicrobial coatings for food packaging papers (40), while a poly(ethylene imine)-modified gelatin nanoparticle was utilized as a biodegradable and incredibly effective protein delivery technology for application in cancer treatment and regenerative medicine (41–43).

Hyaluronic acid and alginate are the most researched polyanions, while chitosan and poly (L-lysine) are typically used as polycations. Polyelectrolyte multilayered coatings such as PLL/HA (44), chitosan/HA (45), and chitosan/ALG (46) have been extensively studied as substrates for cell adhesion. The chitosan/HEP and COL/HEP architectures were excellent candidates to be used as covering film for a variety of implants, including titanium implants for tissue engineering purposes, due to their blood-compatibility qualities (47)].

Additionally, the chitosan/HA coatings showed outstanding adhesive and anti-inflammatory qualities that made them suitable for wound-healing applications (48).

Synthetic polyelectrolytes, such as poly(sodium styrene sulfonate) (PSSNa), poly(acrylic acid), PMMA, polycyclic aromatic hydrocarbons (PAHs), poly(diallyl dimethylammonium chloride) (PDADMAC), poly(ether imide), and poly(dimethyl siloxane), were also used in biomedical applications. The advantage of using this type of polyelectrolyte is the facility of adjusting some parameters such as pH, ionic strength, thickness, ionic concentration and adhesion (49, 50).

One of the most studied polyelectrolyte multilayers used in biomedical applications is PSSNa/PAH films due to their very good adhesion and proliferation of fibroblasts, endothelial and osteoblastic cells (51–53).

8.8 Drug Delivery

8.8.1 *Poly(lactic acid)*

An overview of poly(lactic acid) in clinical applications has been presented (38). Some PLA composites and copolymers are used for tissue engineering and drug delivery applications.

Over the past few decades, PLA has found many applications in medicine, in particular, as orthopedic implants (38).

8.8.2 *Carrageenan*

Naturally derived polymers are of great interest mainly due to their inherent bioactivity. Derived from red seaweeds, carrageenan is a naturally occurring polysaccharide that has shown promise as a biopolymer for various biomedical applications (54).

Carrageenan possesses unique characteristics, including antiviral, immunomodulatory, anticoagulant, antioxidant, and anticancer properties, making it an appealing candidate for tissue engineering and drug delivery research. The versatile properties of carrageenan and the chemical modifications applied to it have been presented in a review (54).

Due to the importance and interesting properties of carrageenans, various review articles have focused on carrageenan-based structures for different applications in tissue engineering and regenerative medicine (55–57).

In addition, some of the research that takes advantage of carrageenan to formulate and fabricate scaffolds and/or drug delivery systems with high potential for tissue repair and disease curing are shown (54).

8.9 Self-Healing Materials

Vitrimers are a class of plastics which are derived from thermosetting polymers and are very similar to them (58). Vitrimers consist of molecular, covalent networks, which can change their topology by thermally activated bond-exchange reactions. At high temperatures they can flow like viscoelastic liquids; at low temperatures the bond-exchange reactions are immeasurably slow and the vitrimers behave like classical thermosets at this point. Vitrimers are strong glass formers. Their behavior opens new possibilities in the application of thermosets as a self-healing material or for simple processibility in a wide temperature range (58).

Besides epoxy resins based on diglycidyl ether of bisphenol A, other polymer networks have been used to produce vitrimers, such as aromatic polyesters, poly(lactic acid), poly(hydroxyurethane)s, epoxidized soybean oil with citric acid and poly(butadiene) (58).

The manufacture of antibacterial carbon fiber-reinforced plastics (CFRP) for medical applications using a imine-based epoxy vitrimer has been described (59). A liquid curing agent was prepared to include an imine group in the matrix, and was synthesized without a simple mixing reaction and a purification process. The vitrimer used as the matrix for CFRP was prepared by reacting a commercial epoxy matter with a synthesized curing agent. The structural and thermal properties of the vitrimer were determined by Fourier transform infrared spectroscopy, differential scanning calorimetry and thermogravimetric analysis.

In addition, the temperature-dependent behavior of the vitrimer was characterized by stress relaxation, reshaping, and shape memory experiments. The mechanical properties of composites fabricated using vitrimer were fully analyzed by tensile, flexural, short-beam strength, and Izod impact tests and had mechanical properties similar to reference material.

Both the vitrimer and the vitrimer composites showed an excellent antibacterial activity against *Staphylococcus aureus* and *Escherichia coli* due to the imine group inside the vitrimer. Therefore, vitrimer composites have a potential for applications requiring antimicrobial properties, such as medical devices (59).

8.10 Surgical Instruments

8.10.1 Tribological Properties of Polymers in Medical Devices

Within the medical devices industry, functionality and safety are the subject of continuous developments, and advances in polymer engineering continue to lead to materials having enhanced properties, such as reduced density, enhanced strength, improved thermal behavior, transparency, wettability, biocompatibility and chemical resistance (60).

In many instances, polymer materials are competitive alternatives to the use of metals or ceramics (61).

The use of polymers is increasingly prevalent for tribological applications in medical devices, and this has been driven both by requirements related to product shelf life and component functionality.

A literature review on tribological influences associated with biomedical devices, including artificial joints, fracture fixation, and surgical instruments, has been presented (62).

Surface modification techniques can alter the texture of a contact surface through mechanical or chemical processes, and modifications can be in the form of material transfer, coatings or material loss, which will affect the subsequent wear mechanisms and rates.

Surface texturing is also used to provide antibacterial or antiviral properties to a medical device and to control its surface wettability (63–65).

8.11 Microplastics

Microplastics are pervasive in the world. Micro- and nanoplastics have effects on human health when disposed of in the environment. The most toxic and dangerous chemical substances that are contained in all plastic products were described in a study (66).

Microplastic particles have been ubiquitously detected in a broad range of shapes, polymers, sizes and concentrations in the environments of marine water, freshwater (67), agro-ecosystems (68), atmosphere (69), food (70), drinking water (71), and biota (72) in remote locations (73).

Microplastics can contain two types of chemicals (66):

1. Additives and polymeric raw materials (e.g., monomers or oligomers) originating from the plastics, and
2. Chemicals absorbed from the surrounding ambience.

Additives are chemicals that are intentionally added during the plastic production to give plastic qualities like color and transparency and to enhance the performance of plastic products to improve both the resistance to degradation by ozone, temperature, light radiation, mold, bacteria and humidity, and mechanical, thermal and electrical resistance (74).

They include inert or reinforcing fillers, plasticizers, antioxidants, UV stabilizers, lubricants, dyes and flame-retardants. Among the charges used are wood and rock flour, clay, kaolin, graphite, glass fibers, cotton flakes, jute or linen, cellulose pulp, etc.

Inert fillers are materials that are used to modify the strength, working and flow properties, and shrinkage of plastics, while the reinforcing ones, also called fillers, are defined as those with some strength properties that are significantly superior to those of the base resin. These fillers, such as carbon black in rubber, which are mixed in with the polymer, result in an interface volume that is generated at the filler-resin contact surface (66).

8.12 Future Trends in Summary

The use of plastics in medical devices was detailed in this chapter. Here, the properties, requirements, and applications were presented along with a comprehensive overview of the main types of plastics used in medical device applications.

Plastics are mainly used in:

Optical applications, dental applications, orthopedic applications, and surgery. In the field of biomedical applications, they are used in self-healing material and drug delivery.

References

1. V.R. Sastri, *Plastics in Medical Devices: Properties, Requirements, and Applications*, William Andrew, (2021).
2. J.K. Fink, *Contact Lenses: Chemicals, Methods, and Applications*, John Wiley & Sons, (2022).
3. L. Nadareishvili, N. Lekishvili, G. Zaikov, and L. Khananashvili, *Polymers and Polymeric Materials for Fiber and Gradient Optics*, CRC Press, (2023).
4. N. Efron, *Contact Lens Practice-E-Book*, Elsevier Health Sciences, (2023).
5. Contact Lens Spectrum PentaVision, Contact lens spectrum's annual report of major corporate and product developments and events in the

contact lens industry in 2004, as well as predictions for 2005, Technical report, BroadcastMed, South Easton Road, Doylestown, PA, (2004).
6. J.J. Nichols and D. Fisher, Contact lenses 2008, *Contact Lens Spectrum* 24–32 (2009).
7. A. Philips and L. Speedwell, Rigid gas-permeable corneal lens fitting. contact lenses.
8. C.S.A. Musgrave and F. Fang, Contact lens materials: A materials science perspective, *Materials* 261 (2019).
9. J. Etzkorn and B. Amirparviz, Encapsulated electronics, US Patent 8 874 182, assigned to Google LLC, October 28, 2014.
10. R.D. Blum, J.N. Haddock, W. Kokonaski, A.V. Heugten, and J. Hunkeler, Ophthalmic dynamic aperture, US Patent 8 215 770, assigned to E-A Ophthalmics, July 10, 2012.
11. R. Marullo, V. Tran, S. Tao, A. Back, H. Le, and E. Irma, Silicone elastomer-silicone hydrogel hybrid contact lenses, US Patent 20 170 307 779, assigned to CooperVision International Holding Co LP, October 26, 2017.
12. A. Alli and D. Sinha, Silicone hydrogels comprising *N*-alkyl methacrylamides and contact lenses made thereof, US Patent 11 125 916, assigned to Johnson & Johnson Vision Care Inc, September 21, 2021.
13. H.-O. Larsen, Preparation of hydrogel for soft contact lens with water displaceable boric acid ester, US Patent 4 495 313, assigned to Mia Lens Production AS, January 22, 1985.
14. H.O. Larsen and T. Kindt-Larsen, Method of forming shaped hydrogel articles, US Patent 4 680 336, assigned to Vistakon Inc, July 14, 1987.
15. T. Kindt-Larsen, J.C. Heaton, and E.C. Rastrelli, Method of forming shaped hydrogel articles including contact lenses, US Patent 4 889 664, assigned to Vistakon Inc, December 26, 1989.
16. T. Kindt-Larsen, J.C. Heaton, E.C. Rastrelli, and G.A. Hill, Method of forming shaped hydrogel articles including contact lenses, US Patent 5 039 459, assigned to Johnson & Johnson Vision Products Inc, August 13, 1991.
17. H. ElHawary, A. Gorgy, and J.E. Janis, Large language models in academic plastic surgery: The way forward, *Plastic and Reconstructive Surgery Global Open* (2023).
18. J.Q. Tay, ChatGPT and the future of plastic surgery research: Evolutionary tool or revolutionary force in academic publishing?, *European Journal of Plastic Surgery* 1–2 (2023).
19. D. Najafali, E. Reiche, J.M. Camacho, S.D. Morrison, and A.H. Dorafshar, Let's chat about chatbots: Additional thoughts on chatgpt and its role in plastic surgery along with its ability to perform systematic reviews, *Aesthetic Surgery Journal* NP591–NP592 (2023).
20. P. Yang, Y. Ju, Y. Hu, X. Xie, B. Fang, and L. Lei, Emerging 3D bioprinting applications in plastic surgery, *Biomaterials Research* 1–27 (2023).

21. N.E. Vrana, S. Gupta, K. Mitra, A.A. Rizvanov, V.V. Solovyeva, E. Antmen, M. Salehi, A. Ehterami, L. Pourchet, J. Barthes, et al., From 3D printing to 3D bioprinting: The material properties of polymeric material and its derived bioink for achieving tissue specific architectures, *Cell and Tissue Banking* 1–24 (2022).
22. X. Cui, T. Boland, D.D. D'Lima, and M.K. Lotz, Thermal inkjet printing in tissue engineering and regenerative medicine, *Recent Patents on Drug Delivery & Formulation* 149–155 (2012).
23. H.-G. Yi, H. Kim, J. Kwon, Y.-J. Choi, J. Jang, and D.-W. Cho, Application of 3D bioprinting in the prevention and the therapy for human diseases, *Signal Transduction and Targeted Therapy* 177 (2021).
24. J. Jang, J.Y. Park, G. Gao, and D.-W. Cho, Biomaterials-based 3d cell printing for next-generation therapeutics and diagnostics, *Biomaterials* 88–106 (2018).
25. R. Raman, B. Bhaduri, M. Mir, A. Shkumatov, M.K. Lee, G. Popescu, H. Kong, and R. Bashir, High-resolution projection microstereolithography for patterning of neovasculature, *Advanced Healthcare Materials* 610–619 (2016).
26. K. Yue, G. Trujillo-de Santiago, M.M. Alvarez, A. Tamayol, N. Annabi, and A. Khademhosseini, Synthesis, properties, and biomedical applications of gelatin methacryloyl (gelma) hydrogels, *Biomaterials* 254–271 (2015).
27. V. Keriquel, H. Oliveira, M. Rémy, S. Ziane, S. Delmond, B. Rousseau, S. Rey, S. Catros, J. Amédée, F. Guillemot, et al., In situ printing of mesenchymal stromal cells, by laser-assisted bioprinting, for in vivo bone regeneration applications, *Scientific Reports* 1778 (2017).
28. F. Guillemot, A. Souquet, S. Catros, and B. Guillotin, Laser-assisted cell printing: principle, physical parameters versus cell fate and perspectives in tissue engineering, *Nanomedicine* 507–515 (2010).
29. R.J. Mondschein, A. Kanitkar, C.B. Williams, S.S. Verbridge, and T.E. Long, Polymer structure-property requirements for stereolithographic 3D printing of soft tissue engineering scaffolds, *Biomaterials* 170–188 (2017).
30. X. Wang, D. Wu, W. Li, and L. Yang, Emerging biomaterials for reproductive medicine, *Engineered Regeneration* 230–245 (2021).
31. J. Adhikari, A. Roy, A. Das, M. Ghosh, S. Thomas, A. Sinha, J. Kim, and P. Saha, Effects of processing parameters of 3D bioprinting on the cellular activity of bioinks, *Macromolecular Bioscience* 2000179 (2021).
32. M. Askari, M.A. Naniz, M. Kouhi, A. Saberi, A. Zolfagharian, and M. Bodaghi, Recent progress in extrusion 3D bioprinting of hydrogel biomaterials for tissue regeneration: A comprehensive review with focus on advanced fabrication techniques, *Biomaterials Science* 535–573 (2021).

33. Z. Xia, S. Jin, and K. Ye, Tissue and organ 3D bioprinting, *SLAS Technology: Translating Life Sciences Innovation* 301–314 (2018).
34. I.T. Ozbolat and M. Hospodiuk, Current advances and future perspectives in extrusion-based bioprinting, *Biomaterials* 321–343 (2016).
35. E.S. Bishop, S. Mostafa, M. Pakvasa, H.H. Luu, M.J. Lee, J.M. Wolf, G.A. Ameer, T.-C. He, and R.R. Reid, 3-D bioprinting technologies in tissue engineering and regenerative medicine: Current and future trends, *Genes & Diseases* 185–195 (2017).
36. F.D. Al-Shalawi, M.A. Hanim, M. Ariffin, C.L.S. Kim, D. Brabazon, R. Calin, and M.O. Al-Osaimi, Biodegradable synthetic polymer in orthopaedic application: A review, *Materials Today: Proceedings* (2023).
37. V.K. Bommala, M.G. Krishna, and C.T. Rao, Magnesium matrix composites for biomedical applications: A review, *Journal of Magnesium and Alloys* 72–79 (2019).
38. S. Suzuki and Y. Ikada, *Medical Applications* chapter 25, 581–604. John Wiley & Sons, Ltd, (2022).
39. J.M. Silva, R.L. Reis, and J.F. Mano, Biomimetic extracellular environment based on natural origin polyelectrolyte multilayers, *Small* 4308–4342 (2016).
40. P. Nechita and M. Roman, Review on polysaccharides used in coatings for food packaging papers, *Coatings* 566 (2020).
41. M.-J. Chou, H.-Y. Yu, J.-C. Hsia, Y.-H. Chen, T.-T. Hung, H.-M. Chao, E. Chern, and Y.-Y. Huang, Highly efficient intracellular protein delivery by cationic polyethyleneimine-modified gelatin nanoparticles, *Materials* 301 (2018).
42. S.-F. Lin, P.-L. Jiang, J.-S. Tsai, Y.-Y. Huang, S.-Y. Lin, J.-H. Lin, and D.-Z. Liu, Surface assembly of poly (I: C) on polyethyleneimine-modified gelatin nanoparticles as immunostimulatory carriers for mucosal antigen delivery, *Journal of Biomedical Materials Research Part B: Applied Biomaterials* 1228–1237 (2019).
43. K. Zwiorek, J. Kloeckner, E. Wagner, and C. Coester, Gelatin nanoparticles as a new and simple gene delivery system, *J. Pharm. Pharm. Sci* 22–28 (2004).
44. T. Crouzier and C. Picart, Ion pairing and hydration in polyelectrolyte multilayer films containing polysaccharides, *Biomacromolecules* 433–442 (2009).
45. J. Almodóvar, L.W. Place, J. Gogolski, K. Erickson, and M.J. Kipper, Layer-by-layer assembly of polysaccharide-based polyelectrolyte multilayers: A spectroscopic study of hydrophilicity, composition, and ion pairing, *Biomacromolecules* 2755–2765 (2011).
46. K. Park, D. Choi, and J. Hong, Nanostructured polymer thin films fabricated with brush-based layer-by-layer self-assembly for site-selective construction and drug release, *Scientific Reports* 3365 (2018).

47. M. Criado-Gonzalez, C. Mijangos, and R. Hernández, Polyelectrolyte multilayer films based on natural polymers: From fundamentals to bio-applications, *Polymers* 2254 (2021).
48. L. Richert, P. Lavalle, E. Payan, X.Z. Shu, G.D. Prestwich, J.-F. Stoltz, P. Schaaf, J.-C. Voegel, and C. Picart, Layer-by-layer buildup of polysaccharide films: Physical chemistry and cellular adhesion aspects, *Langmuir* 448–458 (2004).
49. R.R. Costa and J.F. Mano, Polyelectrolyte multilayered assemblies in biomedical technologies, *Chemical Society Reviews* 3453–3479 (2014).
50. V. Gribova, R. Auzely-Velty, and C. Picart, Polyelectrolyte multilayer assemblies on materials surfaces: From cell adhesion to tissue engineering, *Chemistry of Materials* 854–869 (2012).
51. Z. Feldötö, M. Lundin, S. Braesch-Andersen, and E. Blomberg, Adsorption of IgG on/in a PAH/PSS multilayer film: Layer structure and cell response, *Journal of Colloid and Interface Science* 31–37 (2011).
52. P. Viswanathan, Y.J. Kim, and J.D. Hong, Nanoporous silver submicrocubes layer-by-layer encapsulated with polyelectrolyte films: Nonenzymatic catalysis for glucose monitoring, *Langmuir* 3452–3460 (2020).
53. Q. An, Y. Zhou, Y. Zhang, Y. Zhang, and F. Shi, A facile method for the fabrication of covalently linked PAH/PSS layer-by-layer films, *RSC Advances* 5683–5688 (2014).
54. A. Jafari, M. Farahani, M. Sedighi, N. Rabiee, and H. Savoji, Carrageenans for tissue engineering and regenerative medicine applications: A review, *Carbohydrate Polymers* 119045 (2022).
55. H. Mokhtari, S. Tavakoli, F. Safarpour, M. Kharaziha, H.R. Bakhsheshi-Rad, S. Ramakrishna, and F. Berto, Recent advances in chemically-modified and hybrid carrageenan-based platforms for drug delivery, wound healing, and tissue engineering, *Polymers* 1744 (2021).
56. E.-M. Pacheco-Quito, R. Ruiz-Caro, and M.-D. Veiga, Carrageenan: Drug delivery systems and other biomedical applications, *Marine Drugs* 583 (2020).
57. R. Yegappan, V. Selvaprithiviraj, S. Amirthalingam, and R. Jayakumar, Carrageenan based hydrogels for drug delivery, tissue engineering and wound healing, *Carbohydrate Polymers* 385–400 (2018).
58. Wikipedia contributors, Vitrimers — Wikipedia, the free encyclopedia, https://en.wikipedia.org/wiki/Vitrimers, (2023). [Online; accessed 19-July-2023].
59. W. Kim, Y.M. Kim, S. Song, E. Kim, D.-G. Kim, Y.C. Jung, W.-R. Yu, W. Na, and Y.-S. Choi, Manufacture of antibacterial carbon fiber-reinforced plastics (CFRP) using imine-based epoxy vitrimer for medical application, *Heliyon* (2023).

60. I. Evangelista, D. Wencel, S. Beguin, N. Zhang, and M.D. Gilchrist, Influence of surface texturing on the dry tribological properties of polymers in medical devices, *Polymers* 2858 (2023).
61. V.R. Sastri, Materials used in medical devices, *Plastics in Medical Devices* 19–31 (2014).
62. X. Zhang, Y. Zhang, and Z. Jin, A review of the bio-tribology of medical devices, *Friction* 4–30 (2022).
63. A. Riveiro, A.L. Maçon, J. del Val, R. Comesaña, and J. Pou, Laser surface texturing of polymers for biomedical applications, *Frontiers in Physics* 16 (2018).
64. Y. Zhou, N.F. Fletcher, N. Zhang, J. Hassan, and M.D. Gilchrist, Enhancement of antiviral effect of plastic film against SARS-CoV-2: Combining nanomaterials and nanopatterns with scalability for mass manufacturing, *Nano Letters* 10149–10156 (2021).
65. O. Velgosova, L. Mačák, E. Múdra, M. Vojtko, and M. Lisnichuk, Preparation, structure, and properties of PVA–AgNPs nanocomposites, *Polymers* 379 (2023).
66. C. Campanale, C. Massarelli, I. Savino, V. Locaputo, and V.F. Uricchio, A detailed review study on potential effects of microplastics and additives of concern on human health, *International Journal of Environmental Research and Public Health* 1212 (2020).
67. C. Campanale, F. Stock, C. Massarelli, C. Kochleus, G. Bagnuolo, G. Reifferscheid, and V.F. Uricchio, Microplastics and their possible sources: The example of Ofanto river in southeast Italy, *Environmental Pollution* 113284 (2020).
68. M.C. Rillig, R. Ingraffia, and A.A. de Souza Machado, Microplastic incorporation into soil in agroecosystems, *Frontiers in Plant Science* 1805 (2017).
69. J.C. Prata, Airborne microplastics: consequences to human health?, *Environmental Pollution* 115–126 (2018).
70. R.H. Waring, R. Harris, and S. Mitchell, Plastic contamination of the food chain: A threat to human health?, *Maturitas* 64–68 (2018).
71. M. Pivokonsky, L. Cermakova, K. Novotna, P. Peer, T. Cajthaml, and V. Janda, Occurrence of microplastics in raw and treated drinking water, *Science of the Total Environment* 1644–1651 (2018).
72. S. Rezania, J. Park, M.F.M. Din, S.M. Taib, A. Talaiekhozani, K.K. Yadav, and H. Kamyab, Microplastics pollution in different aquatic environments and biota: A review of recent studies, *Marine Pollution Bulletin* 191–208 (2018).
73. T. Cook, How are microplastics transported to polar regions, *Eos* (2019).
74. J.N. Hahladakis, C.A. Velis, R. Weber, E. Iacovidou, and P. Purnell, An overview of chemical additives present in plastics: Migration,

release, fate and environmental impact during their use, disposal and recycling, *Journal of Hazardous Materials* 179–199 (2018).

9

Restoration

9.1 Deterioration of Cultural Heritage

Cultural heritage includes three main categories: tangible (historic buildings, monuments, books, paintings and archaeological objects), intangible and natural heritage (1). The deterioration of cultural heritage over the past few decades has been attributed to rising air pollution, climate change and excessive biological activity.

The degradation of cultural heritage is an undesirable process brought on by several elements such as humidity, pollution, heat and light that contribute to material degradation (2, 2–4).

9.1.1 Polymers Usage

Humanity has benefitted from polymers for a long time, but did not fully understand them until almost the end of World War II. Very few materials were accessible regarding the creation of the items required for life. For the majority of the construction, materials, including steel, glass, wood, stone, brick and concrete, were employed. Cotton, wood, jute and a few other agricultural items were used to make garments and other textiles.

The introduction of new materials is determined by the sharp increase in demand for produced goods. These new materials known as polymers have had a nearly unfathomable impact on how we live today. Epoxy glue, polymer paints, silicon heart valves, polyethylene cups, fiberglass, nylon bearings, plastic bags, and Teflon-coated cookware are just a few examples of the usual products manufactured from polymers and the list is practically endless (5).

Both natural and synthetic polymers play a significant role in the comfort and ease of human life. They are in charge of sustaining life, as well as for things like food, transportation, communication, irrigation, containers, clothes, historical recording, structures and roadways. In this age, it is impossible to envision a human society without synthetic and organic polymers (6–8).

9.2 Science

Science has a significant role in offering solutions to the pressing issues of food, clean, abundant water, air, energy and health in our rapidly technologically developing society. Understanding polymers and their related research studies can help us understand them better in our daily lives by giving us information and new perspectives. Understanding polymers is made possible by the data gathered from the fundamental science courses. These data cover notions from science that are theoretical, factual and functional. It is helpful for individuals who wish to build a career in law, medical industries, business and other fields (5).

In the recent decades, science has provided innovative materials and methodologies for the examination, maintenance, and conservation of artwork, thus solving the problems of deterioration. However, the degradation of historical objects is an inevitable fact that requires the continuous development of advanced intelligent materials capable of counteracting the specific processes of degradation. Thus, an important challenge in research is the systemic formulation of new functional materials used to protect, conserve or restore art objects (9).

The most recent trends in the area of inorganic and organic materials for the conservation of different types of cultural heritage objects were recently reviewed (10, 11).

9.2.1 Polyelectrolytes

Among the several polymer classes, the polyelectrolytes occupy a particular position. Polyelectrolytes are polymers possessing many ionizable groups. They could contain polycations or polyanions. When dissolved in polar solvents, polyelectrolytes are macromolecules that contain a significant number of covalently bonded ionic groups. In general, different types of such groups may be present in polyelectrolytes. Solely one type of charged group, such as only carboxylate groups, involves homogeneous polyelectrolytes. We refer to a molecule as a polyampholyte when both positive (cationic) and negative (anionic) groups are present. Self-assembled structures, like linear protein assemblies or micelles, frequently have a lot of charged groups and may have characteristics that are very similar to those of polyelectrolytes. Developing upon the self-assembly properties of polyelectrolytes, Decher was the first scientist that developed the layer-by-layer (LbL) aggregation of electrolyte (12).

The applicability of layer-over-layer deposited polyelectrolytes for the purpose of cultural heritage protection was described in detail (1). The state-of-the-art of polyelectrolyte coatings leading to multilayer structures and coatings was also presented.

Several parameters that are influencing the architecture of the resultant films and the self-assembly process, such as the type of polyelectrolytes and the sequence of assembly were detailed.

The recent developments in polyelectrolyte coatings have seen a focus on surface modification by polymers in an effort to expand the known adaptability of bulk polymeric materials to films and ultrathin coatings. In recent years, the self-organization of polyelectrolytes, such as polysaccharides, heterocyclic aromatic compounds, clays, dyes, proteins/enzymes, carbon nanotubes, graphene oxide (GO) and other inorganics, has been exploited more and more for the creation of well-defined surfaces and interfaces. With such techniques, multilayer coatings are formed spontaneously on substrates, due to the interactions between the substrate and polyelectrolyte (13).

By repeating the deposition process, uniform and homogeneous coatings can be obtained with a controlled thickness that varies from a few angstroms to a few microns (14).

9.2.2 *Microbial Attack*

The consolidation and protection of paintings, stone, wood, paper, glass, ceramic, and bones is possible by the simple application of organic solutions (through immersion, injection or brushing) of acrylic, vinylic, and silicone-based polymers and/or their copolymers (15). Most of these polymers are also considered to be reversible because, if necessary, the solvent used for the primary application can be used later for the removal of these resins.

However, although the application of these synthetic polymers slows down the deterioration processes, it does not stop them. Therefore, it is important to consider the chemical and physical properties of polymers in order to evaluate their suitability for use in restoration treatments.

The synthetic polymers that are used to protect artworks from the deterioration process can be colonized by the fungi and bacteria responsible for the biodeterioration process (15).

The susceptibility of synthetic poly(acrylic) resins and poly(urethane) resins to microorganisms (*Aspergillus niger* ATCC 9642, *Aureobasidium pullulans* ATCC 15233, *Chaetomium globosum* ATCC 6205, *Cladosporium cladosporioides* ATCC 16022, *Alternaria alternata* BC01, *Penicillium citrinum* LS1 and *Pseudomonas aeruginosa* ATCC 9027) was investigated.

Resins with a solution of 15% 2-*n*-octlyl-4-isothiazolin-3-one, isothiazolinones are used in occupational and industrial applications for their bacteriostatic and fungistatic activity.

The microbial attack was simulated alone and with a biocide and the related growth was observed up to 21 *d* for bacteria and 28 *d* for fungi. The poly(acrylic) and poly(urethane) resins were subjected to microbial attack, regardless of the biocide treatment, with a fungal growth from 60% to the complete coverage of the plate surface.

Penicillium citrinum showed the greatest adaptation ability and was found in all the examined resins. *P. aeruginosa* was visible in all the different resins, regardless of the presence of biocide. An environmental scanning electron microscopic study revealed the presence of fungal conidia and hyphae in the inoculated resins. Fourier transform infrared spectroscopy indicated chemical transformations in the IR spectra, particularly the hydrolysis of esters, with some differences between the resins and the poly(urethane) resins, which were probably due to their different chemical features.

So, the chemical, physical and biological deterioration caused by microorganisms capable of degrading synthetic polymers is still a problem in art restoration and new strategies must be considered to counteract this phenomenon (15).

9.2.3 *Synthetic Polymers*

One of the reasons for introducing synthetic polymers in conservation treatments was the expectation that these materials would be less prone to chemical, physical and biological deterioration than natural organic products (16).

Synthetic polymers have often been used for the protection and consolidation of cultural heritage. Although it was generally thought that synthetic polymers were not susceptible to deterioration, there are now many papers in the scientific literature demonstrating the opposite. The degradation of synthetic polymers can occur due to chemical, physical and biological factors (16). The traditional way for removing a degraded synthetic polymer is the use of mixtures of solvents that pose some health risks. A method was proposed to select bacteria that are able to remove synthetic polymers from cultural heritage surfaces.

The ability of five bacteria to attack Paraloid B72, the most commonly used polymer in conservation treatments, was evaluated by optical and scanning electron microscopy observations, weight loss measurements, Fourier transform infrared spectroscopy and differential scanning calorimetric analysis.

Although none of the bacteria were able to attack Paraloid B72, the methodology developed can be applied to select other bacteria with this ability (16).

9.2.4 *Removal of Polymers*

Since 1960, it has been popular to employ synthetic polymers to slow down the ongoing deterioration of cultural heritage assets, since it was assumed that they were more resistant than natural materials to microbial colonization. Here, epoxy resins, acrylic polymers, organic silicones and fluoro polymers were the compounds most commonly used for stone conservation.

Some acrylics, polyvinyl acetates and alkyds were used to produce and treat other artworks such as paintings (17).

The application of synthetic polymers has become a routine practice in consolidation and protection of cultural heritage buildings and objects. However, their resistance to microbial colonization is questionable from past experience (18).

Many phenomena, processes and/or activities by microorganisms in synthetic polymers are embraced by the terms biodeterioration, biodegradation and biosusceptibility or bioreceptivity.

Biodeterioration decreases the value of a material of relatively high value, while biodegradation occurs in materials of little or no value, e.g., for the elimination of waste.

In connection with this, the term biodeterioration is closely related to inorganic building materials, mainly stone and masonry from heritage, while biodegradation is often associated with plastic and other synthetic polymeric materials. The terms biosusceptibility or bioreceptivity and biodeterioration are confused in some cases. In addition to the always negative connotations of the biodeterioration concept, the target objectives are the effects of colonization on the material. In contrast, in the case of bioreceptivity, the focus is on the effects of the material on colonization.

The presence of microorganisms on synthetic polymers may result in biofouling, the degradation of the polymer and additives/residual monomers and penetration of biological structures or metabolites like pigments into the synthetic matrix.

Some lipophilic pigments are capable of diffusing into the polymer matrix, leading to an ineffective cleaning of the polymer surface. For some of these synthetic polymers, microorganisms and enzymes involved in their degradation have been described (19, 20)

Acrylics may undergo a microbial degradation in two steps:

1. Enzymatic elimination of side groups, and
2. Cleavage of the C–C backbone.

Biochemical processes, mechanisms and initial pioneering species involved in biodeterioration of other synthetic polymers were reported (21).

Several techniques are now available that can detect the presence of microbial growth on polymeric substrates, such as infrared spectroscopy and laser-induced fluorescence scanning (22).

9.3 Layer-by-Layer Architectures

The layer-by-layer architectures obtained from various natural and synthetic polyelectrolytes are very popular for the protection of different surfaces in many industrial fields, such as biomedicine, packaging, environmental, separation/purification membranes, catalysis, corrosion-resistant films, superhydrophobic, anti-fogging or anti-frosting surfaces, and conservation applications (23–29).

9.3.1 Polymers for Multilayer Architecture

Natural biopolymers can be used as multilayer architecture, c.f. Table 9.1.

Table 9.1 Biopolymers for multilayer architecture (30, 31).

Polymer type	Polymer type	Polymer type
Gelatin	Chitosan	Carrageenan
Alginate	Collagen	Fucoidan
Poly(galacturonic acid)	Heparin	Chondroitin sulfate
Hyaluran		

Feldötö *et al.* demonstrated that monoclonal mouse immunoglobulin G (IgG) could be immobilized on multilayers containing polycyclic aromatic hydrocarbon (PAH) and poly(sodium styrene sulfonate) (PSSNa). They observed that the binding capacity of immobilized IgG on PAH/PSSNa architecture is higher than on a clear surface (32).

An *et al.* reported a facile method for the construction of crosslinked polyelectrolyte multilayers using PSSNa/PAH by post-implantation and subsequent photochem (33).

A PAH/PSSNa layer-by-layer assembly containing nanoporous silver submicrocubes was developed for electrochemical glucose sensing (34).

9.4 Future Trends in Summary

Humanity has has benefitted from polymers for a long time, but did not fully understand them until almost the end of World War II. Very few materials were accessible regarding the creation of the items required for life. For the majority of construction, materials including steel, glass, wood, stone, brick and concrete were employed. Cotton, wood, jute and a few other agricultural items were used to make garments and other textiles.

The degradation of cultural heritage is an undesirable process brought on by several elements, such as humidity, pollution, heat and light, that contribute to material degradation.

References

1. I.C. Gîfu, R. Ianchiș, C.L. Nistor, C. Petcu, I. Fierascu, and R.C. Fierascu, Polyelectrolyte Coatings–A Viable Approach for Cultural Heritage Protection, *Materials* 2873 (2023).
2. K.G. Willis, The use of stated preference methods to value cultural heritage in *Handbook of the Economics of Art and Culture*, Vol. 2 145–181. Elsevier, (2014).
3. G.B. Coelho, H.E. Silva, and F.M. Henriques, Impact of climate change in cultural heritage: From energy consumption to artefacts' conservation and building rehabilitation, *Energy and Buildings* 110250 (2020).
4. F. Di Turo, C. Proietti, A. Screpanti, M.F. Fornasier, I. Cionni, G. Favero, and A. De Marco, Impacts of air pollution on cultural heritage corrosion at European level: What has been achieved and what are the future scenarios, *Environmental Pollution* 586–594 (2016).
5. G.T. Reed, Silicon Photonics: The State of the Art (2008).
6. M.L. Weththimuni, M.B. Chobba, D. Sacchi, M. Messaoud, and M. Licchelli, Durable polymer coatings: A comparative study of PDMS-based nanocomposites as protective coatings for stone materials, *Chemistry* 60–76 (2022).
7. J.L. Sanchez-Salvador, A. Balea, M.C. Monte, C. Negro, and A. Blanco, Chitosan grafted/cross-linked with biodegradable polymers: A review, *International Journal of Biological Macromolecules* 325–343 (2021).
8. S. Begum, N.Y. Yuhana, N.M. Saleh, N.H.N. Kamarudin, and A.B. Sulong, Review of chitosan composite as a heavy metal adsorbent: Material preparation and properties, *Carbohydrate Polymers* 117613 (2021).

9. M. Baglioni, G. Poggi, D. Chelazzi, and P. Baglioni, Advanced materials in cultural heritage conservation, *Molecules* 3967 (2021).
10. T. Fistos, I. Fierascu, and R.C. Fierascu, Recent developments in the application of inorganic nanomaterials and nanosystems for the protection of cultural heritage organic artifacts, *Nanomaterials* 207 (2022).
11. T. Fistos, I. Fierascu, M. Doni, I.E. Chican, and R.C. Fierascu, A short overview of recent developments in the application of polymeric materials for the conservation of stone cultural heritage elements, *Materials* 6294 (2022).
12. G. Decher, Fuzzy nanoassemblies: Toward layered polymeric multicomposites, *Science* 1232–1237 (1997).
13. P. Bertrand, A. Jonas, A. Laschewsky, and R. Legras, Ultrathin polymer coatings by complexation of polyelectrolytes at interfaces: Suitable materials, structure and properties, *Macromolecular Rapid Communications* 319–348 (2000).
14. J. Kötz, S. Kosmella, and T. Beitz, Self-assembled polyelectrolyte systems, *Progress in Polymer Science* 1199–1232 (2001).
15. R. Campana, L. Sabatini, L. Giorgi, G. Pettinari, L. Valentini, and P. Gobbi, A multidisciplinary approach in examining the susceptibility to microbial attack of polyacrylic and polyurethane resins used in art restoration, *International Journal of Molecular Sciences* 11725 (2022).
16. F. Troiano, S. Vicini, E. Gioventù, P.F. Lorenzi, C.M. Improta, and F. Cappitelli, A methodology to select bacteria able to remove synthetic polymers, *Polymer Degradation and Stability* 321–327 (2014).
17. F. Cappitelli, P. Principi, R. Pedrazzani, L. Toniolo, and C. Sorlini, Bacterial and fungal deterioration of the Milan Cathedral marble treated with protective synthetic resins, *Science of the Total Environment* 172–181 (2007).
18. J.-D. Gu, Microbiological deterioration and degradation of synthetic polymeric materials: Recent research advances, *International Biodeterioration & Biodegradation* 69–91 (2003).
19. G. Kurowski, O. Vogt, J. Ogonowski, et al., Paint-degrading microorganisms, *Czasopismo Techniczne* 81–92 (2017).
20. J. Purohit, A. Chattopadhyay, and B. Teli, Metagenomic exploration of plastic degrading microbes for biotechnological application, *Current Genomics* 253–270 (2020).
21. N. Mohanan, Z. Montazer, P.K. Sharma, and D.B. Levin, Microbial and enzymatic degradation of synthetic plastics, *Frontiers in Microbiology* 580709 (2020).
22. F. Cappitelli, F. Villa, and P. Sanmartín, Interactions of microorganisms and synthetic polymers in cultural heritage conservation, *International Biodeterioration & Biodegradation* 105282 (2021).

23. M.A. de Carvalho, P.C. Lazari-Carvalho, I.F. Polonial, J.B. de Souza, and P. Magne, Significance of immediate dentin sealing and flowable resin coating reinforcement for unfilled/lightly filled adhesive systems, *Journal of Esthetic and Restorative Dentistry* 88–98 (2021).
24. V. Vergaro, F. Scarlino, C. Bellomo, R. Rinaldi, D. Vergara, M. Maffia, F. Baldassarre, G. Giannelli, X. Zhang, Y.M. Lvov, et al., Drug-loaded polyelectrolyte microcapsules for sustained targeting of cancer cells, *Advanced Drug Delivery Reviews* 847–864 (2011).
25. W. Stockton and M. Rubner, Molecular-level processing of conjugated polymers. 4. Layer-by-layer manipulation of polyaniline via hydrogen-bonding interactions, *Macromolecules* 2717–2725 (1997).
26. J. Cheung, W. Stockton, and M. Rubner, Molecular-level processing of conjugated polymers. 3. Layer-by-layer manipulation of polyaniline via electrostatic interactions, *Macromolecules* 2712–2716 (1997).
27. R.R. Costa and J.F. Mano, Polyelectrolyte multilayered assemblies in biomedical technologies, *Chemical Society Reviews* 3453–3479 (2014).
28. P. Nechita and M. Roman, Review on polysaccharides used in coatings for food packaging papers, *Coatings* 566 (2020).
29. Q. Li, S. Wang, X. Jin, C. Huang, and Z. Xiang, The application of polysaccharides and their derivatives in pigment, barrier, and functional paper coatings, *Polymers* 1837 (2020).
30. J. Zhang, B. Senger, D. Vautier, C. Picart, P. Schaaf, J.-C. Voegel, and P. Lavalle, Natural polyelectrolyte films based on layer-by layer deposition of collagen and hyaluronic acid, *Biomaterials* 3353–3361 (2005).
31. J.M. Silva, R.L. Reis, and J.F. Mano, Biomimetic extracellular environment based on natural origin polyelectrolyte multilayers, *Small* 4308–4342 (2016).
32. Z. Feldötö, M. Lundin, S. Braesch-Andersen, and E. Blomberg, Adsorption of IgG on/in a PAH/PSS multilayer film: Layer structure and cell response, *Journal of Colloid and Interface Science* 31–37 (2011).
33. Q. An, Y. Zhou, Y. Zhang, Y. Zhang, and F. Shi, A facile method for the fabrication of covalently linked PAH/PSS layer-by-layer films, *RSC Advances* 5683–5688 (2014).
34. P. Viswanathan, Y.J. Kim, and J.D. Hong, Nanoporous silver submicrocubes layer-by-layer encapsulated with polyelectrolyte films: Nonenzymatic catalysis for glucose monitoring, *Langmuir* 3452–3460 (2020).

10

Food Applications

10.1 Molecularly Imprinted Polymers

Molecularly imprinted polymers (MIPs) are bio-inspired synthetic materials capable of being selectively attached to a target molecule.

Since they have outstanding properties for the detection, extraction, quantification, recovery and removal of specific substances, MIPs have found a niche in the preparation of food samples. This has led to the development of novel analytical methods for the efficient detection of trace substances, or even the separation of enantiomers in foods through their use, since they are stable in a wide range of pH, solvents and temperature (1).

These materials are synthesized by using a template molecule, i.e., the same target molecule or a dummy template molecule with a similar structure to the target molecule, a crosslinking agent, a functional monomer and an initiator via different polymerization mechanisms, mainly free-radical polymerization and a reversible deactivation radical polymerization.

In detail, this may be an atom transfer radical polymerization, iniferter polymerization, nitroxide-mediated radical polymerization and reversible addition-fragmentation chain transfer polymerization and methods, e.g., bulk polymerization, emulsion polymerization, Pickering emulsion polymerization, precipitation, surface imprinting, suspension polymerization (1).

Different MIP structures (hydrogels, membranes (electrospun nanofibers and films) and particles (core-shell and hollow-shell nanoparticles, microbeads, nanopillars and nanotubes)) for food applications (sensors for the detection of food contaminants and the quantification of food nutrients and nutraceuticals, active food packaging applications and sample preparation: Removal, preconcentration and detection of target analytes) were comprehensively analyzed based on the literature published from 2016 to 2021 (1).

10.1.1 Extraction of Tetracycline Residues

Tetracycline drugs are mainly used to treat and control diseases and in some cases are used as growth stimulant, particularly in poultry farms. However, the unsafe residue levels of tetracycline in animal-derived foods have raised concerns about food health issues because the residue could lead to gastrointestinal diseases or allergic reactions (2, 3). A tetracycline residue in the environment can lead to an antibiotic resistance (4).

So, many countries have limited the maximum amounts of residue for tetracycline in food samples from animal sources.

Core-shell tetracycline hydrochloride-MIPs were prepared by precipitation polymerization using a magnetic covalent organic framework (Fe_3O_4Tppa-2) synthesized by room-temperature ultrasonic method as a substrate.

The introduction of Fe_3O_4Tppa-2 into the molecular imprinting system can be used to efficiently shorten the time for the adsorption equilibrium of MIPs due to its unique pore structure. The maximum adsorption capacity of MIPs was 87.50 $mg\,g^{-1}$, and the adsorption equilibrium was reached in 14 *min*. The MIPs were used as magnetic solid-phase extraction sorbents in combination with high-performance liquid chromatography for the determination of tetracycline in pork, chicken and chicken liver, and the detection limit reached 0.01 $ng\,ml^{-1}$.

In addition, the adsorption advantages of MIPs suggest that it is a promising adsorbent for the recovery of antibiotics and has potential applications in the detection and enrichment of tetracycline in food.

The recommended maximum residue limit is 100 $ng\,g^{-1}$ for tetracycline in animal-derived food.

The commonly used methods to extract and remove tetracycline from food matrices mainly include solvent extraction, ultrasonic-assisted extraction, dispersive solid-phase extraction, solid-phase extraction, matrix solid-phase dispersion, and microextraction (5).

These methods have the virtues of simplicity, flexibility, and simultaneous extraction of multiple organic analytes (6).

However, they still possess some shortcomings such as high expense, low extraction efficiency, and especially lack of selectivity for specific targets (7).

10.1.2 Mycotoxins

The topic of mycotoxins concerning food safety is becoming increasingly popular in China and worldwide. Mycotoxins are truly involved in a variety of intersecting multidisciplinary directions, such as agriculture, food, and microbiological sciences, as well as analytical chemistry and bioinformatics. An important summary of theoretical and technical advances regarding the detection, risk assessment, and control of mycotoxins contaminating the entire food chain, has been presented in a monograph (8).

As toxic secondary metabolites produced by filamentous fungus, mycotoxins can accumulate in crops (e.g., corn, cereals, soybeans, sorghum, and peanuts) and animal products by entering into the food chain (9–11).

Because of growing concerns about health hazards posed by mycotoxins, reliable analytical methods for the sensitive and efficient detection of mycotoxins in food products are in great demand to properly safeguard human health.

Currently, a number of technologies, including high-performance liquid chromatography and gas chromatography, coupled with effective detection systems such as fluorescence, mass spectrometry, and diode array detectors, have been validated and accepted as recommended tools and/or protocols for the detection of mycotoxins by regulatory authorities. Although conventional methods can accurately detect mycotoxins, their practical or large-scale application is often limited due to the high cost, lack of proficiency in operation, and the laborious sample preparation and analytic processes.

To overcome these limitations, sensors are recommended with many beneficial properties such as high sensitivity, easy detection, cost-effectiveness, rapid response, and abundant signal-transducing mechanisms (such as optical, electrochemical, and mass). Such sensors are also applicable in the field of biomolecular analysis, environmental monitoring, and food safety (12).

The performance of a sensor generally relies on the recognition layer/element. For instance, sensors that use electrochemical techniques without recognition elements suffer from a poor specificity and selectivity towards target analytes. As such, the use of antibodies, enzyme, and aptamers can be considered as recognition elements to make sensors with improved specificity and rapid detection.

However, the natural recognition elements generally suffer from specific storage condition, complicated production process, poor stability, e.g., chemical, thermal, and pH, and high cost. It is thus desirable to build natural recognition element-based sensors that can resolve their drawbacks, especially poor stability and high cost. In an effort to overcome the limitation of such sensing systems, the potential of MIPs has drawn great attention in the construction of sensors due to their promising role as synthetic recognition elements with high specific selectivity and considerable stability.

Molecularly imprinted polymers (MIPs) are synthetic polymers that are produced from functional monomers and crosslinkers in the presence of a specific template for a target analyte molecule.

The cavities within the MIP architectures should perfectly mimic the shape of the target molecules obtained after removing the template with enhanced affinity and specificity

Due to their exclusive advantages (including high selectivity and sensitivity, low cost, and chemical/thermal stability), MIPs have been widely exploited for various sensing applications such as electrochemical, optical, and photoelectrochemical biosensors.

To date, MIP-based sensors have been employed broadly to detect mycotoxins in food samples. However, the mycotoxin sensing has commonly been made with an emphasis on surface-enhanced Raman spectroscopy and/or nanostructured material-based mycotoxins sensors.

In light of the shortage of comprehensive reviews on the MIP-based sensors with respect to the detection of mycotoxins, this review is organized to describe the synthetic strategies for MIPs and their performance (e.g., in terms of quality assurance such as detection limit) based on the various sensing principles of MIPs. Furthermore, ample insights are provided into the current challenges hindering the development of MIP-based sensing techniques for mycotoxin.

Many types of polymerization approaches (e.g., bulk polymerization, sol-gel method, and emulsion polymerization) are used for the preparation of multifunctional MIP particles. Bulk polymerization is used to fabricate MIPs using solution-fabrication after drying (1). The obtained bulk MIPs are further crushed and sieved to reduce binding sites with irregular shapes and different sizes (13).

The high selectivity of MIPs is ascribable to their tailor-made binding sites, which fit with the template molecules (target species) in terms of shape, size, and/or functional groups.

Moreover, stable MIPs can be generated when a strong polymer matrix is formed by the crosslinking components.

As such, the unique characteristics of MIPs (e.g., high selectivity and stability) endow their fabricated mycotoxins sensors with high selectivity and performance evaluation of MIP-based sensors for detection of mycotoxins

The synergistic combination of MIPs with diverse functional materials, listed in Table 10.1, results in composites with sensitive and

Table 10.1 Functional materials.

Abbreviation	Full name
CdTe QDs	CdTe quantum dots
UCNPs	Upconversion Nanoparticles
CNTs	Carbon nanotubes
GQDs	Graphene quantum dots
rGO	Reduced graphene oxide
MOFs	Metal organic frameworks

selective detection of mycotoxins.

Using MIP particles or films, various types of MIP-based sensors have been constructed as either optical sensors (e.g., fluorescence, phosphorescence, SPR, and SERS), electrochemical sensors (e.g., electrochemical and photoelectrochemical), or piezoelectric sensors, (e.g., a quartz crystal microbalance (QCM)-based sensor), for the detection of mycotoxin in various food samples. The sensing capacity of MIP-based sensors against mycotoxins is generally improved by the incorporation of other materials, including noble metals, such as Pt and Au (11).

10.2 Self-Assembled Carbohydrate Polymers

Natural polymers obtained from native or modified carbohydrate polymers, such as celluloses, chitosan, glucans, gums, pectins, and starches, have gained special attention as raw material in the manufacture of self-assembled polymer composite materials having several forms: films, hydrogels, micelles, and particles (14).

The easy manipulation of the architecture of the carbohydrate polymers, as well as their high availability in nature, low cost, and their being sustainable and green polymers, have been the main positive points in their use for different applications. Carbohydrate polymers have been used as building blocks for composite structures, and their easy orientation and ordering has given rise to self-assembled carbohydrate polymers (14).

These macromolecules were studied only a little for food applications. Nonetheless, their research has grown mainly in the last five years as encapsulated food additive wall materials, food coatings, and edible films. The multifaceted properties (systems sensitive to pH, temperature, ionic strength, types of ions, mechanical force, and enzymes) of these devices are leading to the development of advanced food materials (14).

10.3 Quartz Crystal Microbalance Sensors

10.3.1 Fabrication Methods

Several acrylic monomers containing aromatic linkers were synthesized under the assumption that the aromatic groups, in particular the electron-rich groups, would induce favorable electronic interactions (15). These p-interactions are the only directional attractive forces that can be used for molecules like chlorinated aromatics or polyaromatic molecules.

Some of these monomers are:

1. 9,9-Diallyl fluorene
2. Diallyl hydroquinone
3. Tripropyleneglycol diacrylate
4. Trimethylopropane triacrylate

Experiments with simple acrylic networks obtained by photoirradiation of trisacrylate, such as tripropyleneglycol diacrylate or trimethylopropane triacrylate, a reactive and benzyl methacrylate, indicated a very poor adhesion of the film on the metallic surface.

The film peeled off instantaneously after immersion in either dichloromethane or acetonitrile. This behavior was both expected and surprising. It was surprising because claims made in the recent literature imply that conditions can be found to obtain good adhesion of acrylic or styrenic polymers on a gold surface. It was expected as this problem has a long history in polymer science and engineering and is still considered today as one of the key problems in UV-curing on metallic surfaces (UV-cured paintings or plastic coatings on cars are traditional examples).

The physics of the phenomenon is complex in its details, but the rationale for the poor adhesion is well known and understood. The polymer film shrinks during the curing while the underlying metallic substrate does not, generating a very large amount of stress at the interface. Strategies that were successively considered to solve this problem include (15):

1. Use of acrylic acid as a co-monomer,
2. Use of itaconic acid as a co-monomer,
3. Modification of the gold surface with a thiol-capped acrylic monomer able to self-assemble on the surface,
4. Use of tetraethylthiuram disulfide as a photoinitiator or chain-transfer agent,
5. Use of triacrylate as the multifunctional monomer, and
6. Use of bis(dehydroalanine) as the multifunctional monomer.

A schematic representation of a quartz crystal microbalance detector assembly is shown in Figure 10.1.

Here, the ports 40 and 42 in the sensor assembly may be modified so that an inert shrouding gas is flowed through such ports during the period of operation that the gas stream otherwise being sensed contains the undesirable component that in contact with the microbalance sensor coatings would adversely affect such coatings (15).

The quartz microbalance sensor assembly 10 uses two quartz microbalances 12 and 14 whose temperatures are independently regulated in each respective chamber 44 and 46 of housing 34. The

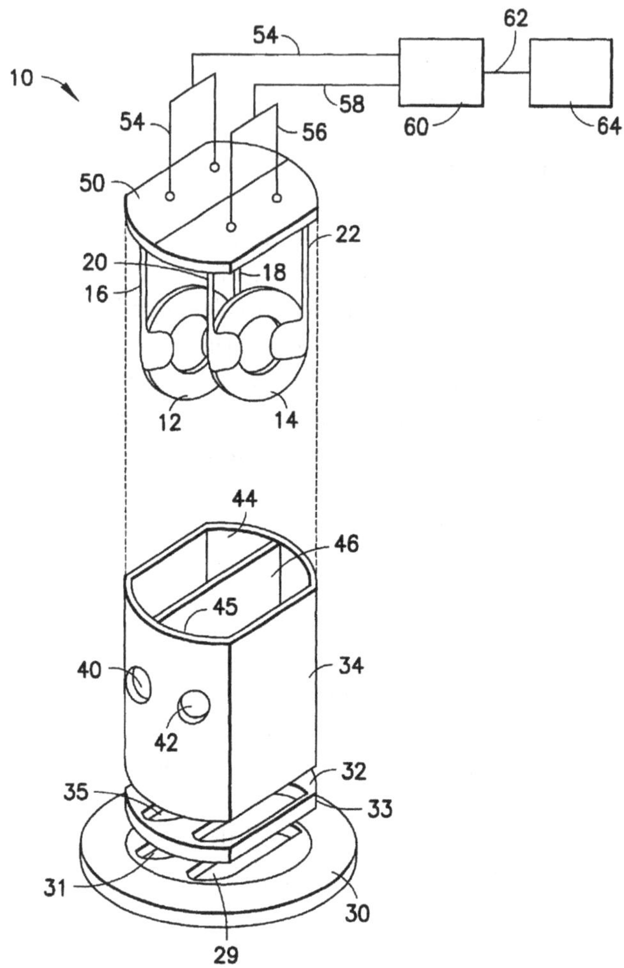

Figure 10.1 Quartz crystal microbalance detector assembly (15).

chambers 44 and 46 are defined in the interior volume of the housing by partition 45. One of the chambers 44 and 46 may be selectively cooled in relation to the other chamber, to make detection of higher vapor pressure gases easier by suitable thermal adjustment and monitoring means (15).

The sensor assembly 10 includes the aforementioned quartz microbalances 12 and 14 mounted on the end plate 50 which cooperatively mates with the housing 34, being bonded, mechanically fastened or otherwise secured thereto, so that the quartz microbalance element 12 is disposed in chamber 44, and the quartz microbalance element 14. The quartz microbalance element 12 is joined by leads 16 and 18 to associated exterior lead assembly 52 and signal transmission line 54 to processing circuitry in the electronics module 60.

Likewise, the quartz microbalance element 14 is joined by the leads 20 and 22 to associated exterior lead assembly 56 and signal transmission line 58 to the processing circuitry in the electronics module 60.

The processing circuitry in the electronics module 60 includes oscillator circuitry and signal processing circuitry for applying an oscillating electric field to the piezoelectric crystal that generates an output resonant frequency therefrom, sampling the output resonant frequency of the piezoelectric crystal while the oscillating electric field is applied thereto, determining the change in resonant frequency, or the rate of change of resonant frequency, incident to the interaction of the sensor material with the gas species of interest, and generating an output indicative of the presence of the gas species of interest in the gas mixture being monitored.

The output from the electronics module is passed by signal transmission line 62 to the process monitoring and feedback module 64. Module 64 may include visual display means such as a monitor, recorder, flashing light or colorimetric alarm, etc., to provide a visually perceivable indication of the presence and/or concentration of a particular gas species in the gas mixture being monitored by the system.

The module 64 may also include audio alarms for output reflecting the presence of a particular gas species in the monitored gas mixture, or other output or display means for such purpose. Further, the module 64 may comprise adjustment or modulating means for adjusting a process or source of the gas mixture, to obtain a desired concentration level of a particular gas species in the multi-component gas mixture, or for effecting a change in the treatment of a gas mixture to abate a particular gas species therein.

For example, the gas mixture may comprise a multi-component gas stream that is flowed into the housing 34 to permit sensing of specific components by the quartz microbalance elements in the respective chambers 44 and 46. In response, an upstream process system producing the gas stream may be moderated by feedback from the module 64, to adjust the process conditions (e.g., pressures, temperatures, flow rates, compositions, etc.) of the upstream process so that it yields a product gas stream of the desired character.

For purposes of monitoring the respective components of a multi-component gas mixture in the housing chambers 44 and 46, and independently adjusting the temperature and pressure of the respective chambers, the housing 34 is provided with pressure, temperature monitoring ports 40 and 42, which accommodate the coupling with the housing of suitable means (not shown) for monitoring the pressure and/or temperature of a chamber, for cooling or heating a chamber for a specific gas component sensing, etc.

The lower end of the housing 34 of the sensor assembly 10 mates with thermal insulating gasket 32 having openings 33 and 35 therein, with the gasket in turn mating with adapting flange 30 having openings 29 and 31 therein, so that the opening 29 of the adapting flange 30 is in register with opening 33 of gasket 32, and the registered openings 29 and 33 are in gas flow communication with the chamber 46 of the housing 34.

Concurrently, the opening 31 of the adapting flange 30 is in register with the opening 35 of the gasket 32, and the registered openings 31 and 35 are in gas flow communication with the chamber 44 of the housing 34. The flange 30 may be connected to a piping fitting, to a conduit or to a process chamber of an industrial facility, to permit the gas mixture to be monitored to permeate through the registered openings to the quartz microbalance elements in the respective chambers for detection of specific gas components therein.

In this manner, the gas mixture being monitored can be sensed in each of the two chambers 44 and 46, e.g., with different sensor coating materials on each of the respective quartz microbalance elements so that different components of the gas mixture are able to be monitored.

Alternatively, the gas mixture can be simultaneously monitored in the respective chambers at different conditions, e.g., different temperature conditions. For this purpose, the partition 45 as well as the gasket 32 may be formed of a thermal insulating material, and the housing 34 itself may be formed of a material having suitably low thermal conductivity. It will be appreciated that the quartz microbalance assembly of the invention may be arranged with more than two quartz microbalance elements, each in a separate chamber, so that more than two components of the gas may be individually monitored, and/or so that the gas mixture may be monitored under a variety of sensing conditions.

Alternatively, the quartz microbalance assembly may be arranged with only a single quartz microbalance element, but with the element being subjected to exposure to different gas flows at different times, e.g., by means of a cycle timer program controlling valves in flow circuitry including different gas flow streams, so that gas is sampled and contacted with the quartz microbalance element at predetermined intervals or in a predetermined sequence.

Accordingly, it will be appreciated that the quartz microbalance assembly of the invention may be alternatively configured, arranged and operated, in a wide variety of differing embodiments (15).

10.3.2 Vapor Deposition

Vapor deposition, in particular physical vapor deposition and chemical vapor deposition, is an umbrella term used to describe a variety of low pressure processes to condense and deposit vaporized material onto workpiece surfaces. Physical vapor deposition involves purely physical processes such as high temperature vacuum evaporation followed by condensation, or plasma sputter bombardment. Chemical vapor deposition employs chemical reactions to produce high-purity, high-performance solid materials (16).

This process is used in semiconductor manufacturing to form thin films. Substrates are exposed to one or more vaporized chemicals which react or decompose on the target surface to produce the desired deposit. Often, volatile by-products are produced. These are removed from the reaction chamber by carrier or reaction gases.

Atomic layer deposition (ALD) is a chemical vapor-based method well suited to form conformal films on both planar and complex structures. Typically, ALD is based on sequential and self-limiting surface reactions, whereby thin film deposition is controlled by surface chemistry at the atomic scale. Typically, ALD requires two chemically selective half-reactions utilizing vapor phase chemical species, which are often an organometallic precursor and a co-reactant such as water, ammonia, or hydrogen sulfide separated in time or space by an inert gas purge.

During the first half-reaction, a first chemical precursor is introduced into the reaction chamber and reacts with surface species, changing the initial surface termination chemistry, perhaps to contain first precursor moieties. This self-limiting aspect prevents further deposition of the first precursor. Following a purge step to remove remnants of the first half-reaction, a second chemical precursor or reactant contacts the surface until all reactive surface sites have been terminated with the new species. The reaction chamber is again purged to remove unreacted precursors and products (16).

The process is repeated if necessary to attain the predetermined thickness of the film desired. One benefit of the self-limiting nature of ALD growth is uniform deposition largely independent of chemical precursor exposure time. This means that even large and long reactant doses, which are necessary to reach and react with the most distant sites on convoluted, high surface area structures, result in conformal and pinhole-free growth without excess buildup of material on any one region. These features, combined with low growth temperatures and medium vacuum processing, make ALD a suitable vehicle to prepare high-quality films of materials such as oxides, nitrides, sulfides, and even pure elements (16).

However, there exists a need for *in-situ* measurements, in particular mass and thickness measurements, to more thoroughly understand the process and rapidly advance the ALD art. Obtaining accurate measurements of tiny changes in mass during ALD is difficult because state-of-the-art microbalances, such as those utilizing oscillating crystal paradigms, are intended to be operated at room temperature.

For example, AT-cut quartz crystals (utilized in quartz crystal microbalances or QCM) are extremely sensitive to temperature fluctuations when operated at temperatures exceeding 50°C, AT-cut crystals, typically used in sensor applications, comprise quartz blanks formed from a thin plate cut at an angle of about 35° 15′ to the optic axis of the crystal. At the intended operating temperature for the AT-cut crystal of 25 Hz/°C, the temperature coefficient is 0 Hz/°C. However, at a typical ALD temperature of about 180°C, the temperature coefficient for the AT-cut quartz crystal is 50 Hz/°C. Therefore, a 1°C temperature increase will result in an apparent change in the ALD Al_2O_3 thickness of minus 17 Angstroms (Å), which represents more than 15 ALD coating cycles.

Tight temperature control, e.g., with proportional, integral and derivative (PID) controllers to reduce temperature-induced variations from the mass balance, may not always be practical. Potentials for error increase with increasing temperature, reaching more than 200 Å/°C under some ALD-relevant conditions. Moreover, thermal transients induced by the ALD precursor exposures cannot be eliminated even using PID temperature controllers.

To minimize the temperature effects of QCM measurements during ALD, another crystal, e.g., a $GaPO_4$ sensor, can be used instead of the traditional AT-cut quartz crystal. Over the range of typical ALD processes (25–400°C) the temperature coefficient of the $GaPO_4$ sensor is on average, much lower. However, the temperature sensitivity of $GaPO_4$ systems at temperatures less than 150°C is greater than AT-cut quartz. Accurately estimating the mass deposited during ALD is further complicated by deposition on both the front and back of the oscillating crystal, which is an inevitable consequence of the self-limiting surface chemistry.

Several shortcomings limit the utility of oscillating crystal systems for ALD (16). First, temperature equilibration of conventional QCM sensor heads (i.e., fixture on stick configurations) requires several hours. This, combined with a significant disturbance of the gas flow dynamics produced by conventional QCM sensor heads, precludes the technique as a routine *in-situ* monitor of thickness for co-deposited samples of interest. Given the growing efforts to develop more effective ALD processes and materials, there exists a need for *in-situ* characterization during regular/normal operation.

This *in-situ* characterization should be seamlessly integrated with the ALD equipment for convenience and high throughput. The measurement system should accommodate variations in temperature and process conditions to provide an accurate measurement of the ALD process, and particularly to provide an accurate characterization of films as they are being fabricated. This will allow film thicknesses to be tuned during their fabrication and also allow films to be characterized at several locations of the film surface simultaneously.

The system should also provide rapid temperature equilibration after substrate loading, and present a minimal disturbance to the ALD environment. Finally, the system should be easily implemented in a broad variety of ALD equipment platforms having different sizes, shapes, and configurations (16).

10.3.3 *Analyte-Responsive Polymers*

Detection mechanisms capable of detecting chemical and biological substances in the nanomolar range are finding increasing utility in a variety of commercial fields, including the food, healthcare, environmental and waste treatment industries (17). A number of sensitive devices have been developed to meet these detection needs, including those taking advantage of shifts in resonant frequency, such as piezoelectric transducers as well as optical sensors which operate on the basis of changes in refractive indices.

The ability of piezoelectric transducers to detect small changes in mass, viscosity and density at their surfaces while immersed in liquids has made them particularly useful as analytical tools where the measurement of very small amounts of material must be made in solution.

Design strategies for piezoelectric sensors have primarily involved events that result in mass changes at the surface of the crystal. The most extensively studied transducer in this regard has been the shear mode AT-cut quartz resonator, commonly referred to as the QCM, which consists of an AT-cut quartz crystal sandwiched between two metal excitation electrodes. An example of the usefulness of the QCM is its use in detecting biological analytes.

This can be seen in a patent from Karube and Muramatsu (18), where the concentration of an analyte in solution was calculated on the basis of the change in resonance frequency, caused by the weight of an analyte added to a receptor material immobilized on the surface of a piezoelectric device. The invention of Karube and

and Muramatsu demonstrated that the piezoelectric transducer was useful in measuring the concentration of some biological analytes.

Although a useful tool in solution environments, a piezoelectric transducer must have its surface modified in order to give it a measure of specificity for the analyte to be detected, and preparation of these modified piezoelectric devices is procedurally complex and often difficult. In cases where the analytes to be detected are biological in nature, receptor agents (antigens, antibodies or other ligands) must be immobilized on the surface of the crystal. However, this method has significant inherent limitations as the receptor agents can inactivate during the immobilization process or separate from the surface of the crystal after immobilization.

To facilitate receptor modification of the QCM, polymer-coated quartz oscillators have been developed that allow for more efficient and specific binding of receptor agents. Mueller-Schulte and Laurs (19) described the coating of a piezoelectric oscillator with a water-insoluble polymer that facilitates the absorbtion of various biomolecules. These immobilized biomolecules can then bind antigens or ligands and the resulting increase in mass of the oscillator quartz translates into changes in resonant frequency which can be used to quantitate the analyte.

These examples of the use of the piezoelectric sensors are useful but limited in that they all rely on a mass change on the sensor surface to effect a change in resonant frequency. Sensor design based solely on mass changes can be limiting if the analytes have low molecular mass. For example, the mass increase associated with the binding of a protein to an active surface of a shear mode AT-cut crystal generally will not be sufficient for a practical frequency response. Other considerations include the rigidity of the bound analyte; for example, Newtonian films that bind to the surface of a QCM give frequency shifts that are much smaller than predicted by the Sauerbrey equation (17).

The Sauerbrey equation is a method for correlating changes in the oscillation frequency of a piezoelectric crystal with the mass deposited on it (20). Sauerbrey simultaneously developed a method for measuring the characteristic frequency and its changes by using the crystal as the frequency-determining component of an oscillator circuit. This method continues to be used as the primary tool in quartz crystal microbalance (QCM) experiments for conversion of frequency to mass and is valid in nearly all applications (20).

Additionally, piezoelectric oscillator techniques have not been well-suited to continuous real-time (kinetic) measurements of the biological activity of organisms and the usefulness of receptor-modified piezoelectric methods has been limited by the specificity of known reagents, a fact that requires repeated testing to assess the presence of different organism types (17).

To overcome these limitations, polymer films have been designed to react with specific reagents, ions or metabolites in an attempt to increase the specificity of the detection method. For example, in an U.S. patent (21) it was disclosed that propylene imine will react with polyampholytes via a ring-opening mechanism to form primary amines. The primary amines can react with tanning developers and with ordinary aldehyde crosslinking agents to form crosslinked networks.

Ebersole *et al.* (22) described the use of polyampholytes in solution which, upon contact with the metabolic products of an organism, deposit upon a piezoelectric device to produce a change in resonance frequency that is correlated to the concentration or rate of change in the presence of a metabolic product. Here, the polyampholyte is not immobilized on the surface of the sensor before exposure to an analyte suspected of being present in a sample. Tanaka and Hirokawa (23) have demonstrated that certain ionic gels formed by the polymerization of isopropylacrylamide in the presence of a metal ion containing monomer, crosslinking agent, and a suitable liquid medium are capable of drastic volume changes in response to changes in solvent composition, temperature and pH or ion composition.

Attempts to integrate the dynamic physical dimensional properties of polymer films into analyte detection systems have been few and poorly developed. Various physical dimensional changes have been reported for crosslinked polymers responding to changes in partial pressures of gases such as oxygen (24), as well as to changes in humidity and salt concentrations.

Although gels and crosslinked polymers are known to be responsive to various biological and ionic analytes as shown in the above art, in all cases the preparations depend on a common physical principle of polymer swelling in response to salt or humidity changes. Furthermore, there is nothing inherent in the polymer gel that makes it specifically responsive to water or a particular ion and all gels must be in a pre-swollen state in order to function.

Optical sensors for the detection of analytes generally rely on small changes in the indices of refraction in response to the presence of an analyte. Commonly used optical sensors include planar waveguides, optical fibers and diffraction gratings. In general, optical sensors of this sort suffer from many of the same disadvantages as piezoelectric oscillators such as lack of specificity for an analyte, difficulty in surface preparation as well as difficulty in measuring continuous real-time (kinetic) changes in biological activity. To date, these difficulties in the field of optical sensors remain unaddressed.

10.3.4 *Specific Materials and Methods of Synthesis*

Specific polymers used as analyte-responsive polymers for the examples of a sensing device using acoustic or light energy are co- or terpolymers of acrylic acid, alkyl methacrylate, and *N,N*-dimethylaminoethyl methacrylate.

Structurally, the polymers may be linear or may contain pendent or crosslinking chains. They are prepared using a two-step process (25).

First, a mixture of the monomers acrylic acid, and *N,N*-dimethyl- or *N,N*-diethylaminoethyl methacrylate and, optionally, esters of methacrylic or acrylic acid, is used for the preparation by emulsion polymerization. The methyl ester of acrylic acid and the other monomers are polymerized to yield an acrylate-containing prepolymer. Second, the acrylate ester groups are selectively hydrolyzed, in the presence of a base, to yield the desired amphoteric polymer with pendant carboxyl and dialkylaminoalkyl ester groups.

This two-step procedure is an improvement in the polymerization of the monomers because the resultant polymer is substantially free of any betaine-type contaminant. Preferably, the polymerization step is carried out by emulsion techniques because the reaction proceeds more rapidly in this manner than by solution techniques. Emulsion polymerization can be carried out by known procedures, preferably using potassium persulfate as the polymerization initiator.

The polymerization temperature is preferably about 50°C to 70°C.

However, with a redox initiator system, e.g., potassium persulfate/*N*,*N*-dimethylaminoethanol, temperatures as low as about 0°C can be used. The polymerization can be accomplished by batch or continuous processes. An expanding batch process with specific, gradual, controlled addition of monomer, initiator and emulsifier solution, and a continuous overflow process are preferred.

During the hydrolysis step, acrylate ester groups are rapidly converted to carboxylate salt groups in the presence of a base. Since this hydrolysis reaction is several orders of magnitude faster than the corresponding hydrolysis of methacrylate ester groups, selective hydrolysis of the acrylate ester groups can be achieved.

If a limiting quantity of base is used, hydrolysis will proceed only until the base is consumed. Hence, the degree of prepolymer hydrolysis and therefore the ratio of carboxyl to dialkylaminoalkyl groups in the amphoteric polymer can be regulated by the relative quantity of base used.

The basic hydrolysis of the prepolymer is preferably carried out with aqueous potassium hydroxide, preferably in about a 10% to 20% solution of the base and preferably at a temperature of about 65°C to 90°C.

A neutralization of the hydrolyzed polymer can be accomplished with a strong acid, e.g., nitric acid, and the polymer can be separated from solution by isoelectric precipitation in excess water. The amphoteric polymer can be redissolved, after purification, at a pH other than the isoelectric point.

Alternatively, polymer neutralization can be accomplished with an acidic ion-exchange resin. Slightly less than the calculated amount of resin is usually employed to facilitate separation of the amphoteric polymer solution which can be used directly, if desired.

10.3.5 Detection of an Analyte in a Liquid Sample

A method has been presented for detecting an analyte in a liquid sample (17). The method uses light energy propagational properties of an analyte-responsive polymer. The analyte-responsive polymer is interfaced with a means to detect changes in the light propagation of the analyte-responsive polymer and the changes are correlated to the amount of analyte in the liquid sample. The method works by (17):

1. Contacting an optical detection system comprising a proton-responsive polymer-coated on the surface of an optical sensor, with an analyte comprising protons to which said polymer is responsive whereby said polymer undergoes propagational changes in response to the presence of protons and wherein said polymer is an amphoteric polymer selected from the group consisting of amphoteric co- or terpolymers of pI between 5.0 to 8.0 of acrylic acid, alkyl methacrylate, and *N,N*-dimethyl-aminoethyl methacrylate.
2. Interfacing the system of step (1) with a means to detect changes in the light propagation of said polymer,
3. Detecting changes in the index of refraction of light at the surface of said polymer coated on an optical sensor whereby said propagational changes are measured, and
4. Correlating the propagational changes measured in step 3 with the presence, concentration, rate of production or consumption of protons.

10.3.6 *Detection of Gases and Humidity*

Graphene and graphene-based materials have a great potential for detecting gases and humidity due to their high specific surface areas. QCM sensors have been massively developed because they have high sensitivity, consume low energy, and can be readily modified. Recently, QCM coated by graphene composites has been explored to sensitively and selectively detect various gases and humidity (26).

The recent progress on QCM gas sensors and QCM humidity sensors based on graphene materials and graphene composites has been detailed (26).

A QCM sensor is a piezoelectric device that is extremely sensitive to the change in mass of a molecule so that the quartz crystal microbalance sensor has high sensitivity and accuracy at room temperature operation (27–33).

The characteristics of this sensor highly depend on the chemical nature and the physical properties of the sensing material coated on the quartz crystal microbalance sensor active electrode.

Furthermore, both mass and viscoelastic properties of the material will affect the response. The viscoelasticity usually has a great influence on surpassing frequency noise during the detection process. By using various kinds of sensing materials, the QCM sensors are widely utilized to selectively detect various targeted materials, c.f. Table 10.2.

Table 10.2 Gas detection.

Material	References
Ammonia	(34, 35)
Alcohol	(32, 33)
Mercury	(36)
Aromatic hydrocarbons	(37)
Hydrogen cyanide	(38)
Trimethylamine	(39)
Dimethyl methylphosphonate	(40)
Humidity	(41)

Some strategies to enhance the selectivity of sensing materials on QCM platforms have been developed, including designing heterostructures, surface functionalization, and fabrication of nanocomposite. On the other hand, the sensitivity of the QCM sensor is improved via increasing the surface area of the sensing material by incorporating the sensing materials with high surface materials such as graphene and its derivatives, i.e., graphene oxide and reduced graphene oxide. Hence, various graphene materials and graphene composites have been utilized as sensing materials for both QCM gases and humidity sensors.

Research on graphene-based QCM sensor applications is currently focused on modifying sensing materials with graphene materials in order to improve the sensor performance, including selectivity, sensitivity, response time and stability. Functionalizing graphene is a good strategy for surface-modified sensing materials because of its high specific surface area and flexibility of tunable surface properties for selective gas sensors. However, the pristine graphene has no dangling bonds on its surface as the main requirement for gas/vapor adsorption in QCM gas sensors. Therefore, GO and rGO are more beneficial than pristine graphene for fabricating graphene-based nanocomposites as sensing materials for QCM sensors since both GO and rGO could be readily functionalized with other materials such as nanoparticles, carbon nanotubes and polymers [68,69].

This review article is intended to focus on the latest advancement and new perspectives of graphene material, including GO, rGO and graphene composites, as sensing materials of QCM sensor for gases and humidity detection. The article is organized into four main parts, i.e., sensing principle of QCM, synthesis and preparation of sensing materials based on graphene materials coated on QCM device, progress on QCM-based gas and humidity sensors using graphene materials and graphene composites as the sensing film, and the sensing mechanism of those sensors. The first part briefly explains the sensing principle of a QCM sensor, including the main factors affecting the quality of QCM sensors. The second part briefly introduces the common synthesis of graphene materials that are commonly used as composites for sensing materials of QCM. In addition, some coating techniques used for QCM sensor preparation are also discussed. The third part mainly focuses on the application of graphene and graphene composites for QCM gas sensors as well as humidity sensors. The fourth part explains the general mechanism of the sensing process of the resultant sensors. Finally, the remaining challenges in the preparation, practical applications and sensing mechanism of QCM sensors coated by graphene materials or graphene composites are also elaborated (26).

10.4 Analysis of Problematic Additives

10.4.1 Pesticide Detection

Molecularly imprinted polymers (MIPs) have added a vital contribution to food quality and safety with the effective extraction of pesticide residues due to their unique properties. Magnetic MIPs are a superior approach to overcome stereotypical limitations due to their unique core-shell and novel composite structure, including high chemothermal stability, rapid extraction, and high selectivity (42).

Over the past two decades, different magnetic molecularly imprinted polymers (MMIPs) have been developed for pesticide extraction in actual food samples with a complex matrix. Such developments are highly desirable. The synthesis and mode of application of MMIP have great potential as a green chemistry approach that can significantly reduce the environmental pollution and minimize resource utilization.

The MMIP application for single or multi-pesticide detection has been summarized by critiquing each method's uniqueness and efficiency in real sample analysis and providing a possible green chemistry exploration procedure for MMIP synthesis and application for escalated food and environmental safety (42).

10.4.2 Aflatoxin in Milk

Aflatoxin M1 is an important risk factor threatening the safety of milk and dairy products due to its carcinogenic and teratogenic effects on humans (43).

To prevent aflatoxin M1 from causing damage to human health, methods to monitor its levels in milk and dairy products were developed. Biosensors built with recognition and detection systems have attracted extensive attention for their simplicity, portability, sensitivity, and selectivity. Commonly developed biosensors for detecting aflatoxin M1 are antibody-based sensors (immunosensors) and aptamer-based sensors (aptasensors).

The advances of immunosensors, aptasensors, and other emerging receptors-based sensors for the detection of aflatoxin M1 in milk and dairy products were described (43). These biosensors were reported with representative examples according to their signal transduction systems, mainly including colorimetry, fluorescence, electrochemistry, and others.

Colorimetric-Based Aflatoxin M1 Immunosensor. The colorimetric signal of a conventional enzyme-linked immunosorbent assay was produced by horseradish peroxidase or alkaline phosphatase by catalyzing colorless organic dyes to produce colored chemical compounds (44).

This conventional signal is limited to single-color response, which makes it unsuitable for accurate naked-eye detection (45, 46). To address this shortcoming, noble metal nanoparticles (NMNPs) were introduced into immunosensors.

A colorimetric aptasensor mainly depends on the target-induced protection of gold nanoparticles against salt-induced aggregation to function (47, 48).

In a typical detecting process, aptamers are adsorbed on the surface of gold nanoparticles to prevent aggregation under a high salt concentration. The targets can cause the dissociation of aptamers from the gold nanoparticles surface due to the higher binding force of aptamer to the targets than to gold nanoparticles. As a result, an aggregation of unprotected gold nanoparticles can be observed.

Antibody and aptamer-based detection methods have been demonstrated to be able to detect targets with desirable selectivity and sensitivity. However, these biomolecular recognition elements require strict testing conditions, such as precise temperature, precise pH values, and the absence of protease/nuclease, to perform well (49, 50).

10.5 Food Packaging

Active and intelligent food packaging is an innovative technology to prevent food contamination and ensure food quality and safety (51). Active packaging protects the food from microbial contamination, while smart or intelligent packaging enables monitoring the freshness of the food or quality change in real time.

10.5.1 Packaging Methods

As the food packaging industry evolves, efforts to ensure sustainable and efficient food quality and safety during transportation and storage continue to grow. Various packaging approaches are presented below.

10.5.1.1 Active Food Packaging Systems

Today, active and intelligent packaging systems are at the forefront of cutting-edge technologies for the food supply chain (52, 53).

An active food packaging film is a preformed, thin-layer, freestanding film with functional properties that can be used in food packaging, driven by the consumer demand for high-quality, safe, natural foods.

As one of the most important active packaging systems, antibacterial packaging protects food from microbial contamination, the most common cause of food spoilage (54, 55). An active packaging film also acts as a UV-blocking material. This helps to prevent an UV-mediated oxidation of foods.

Also, it helps to regulate the gas exchange between fresh products and the environment.

10.5.1.2 *Intelligent Food Packaging Systems*

Intelligent packaging systems are a new technology that is widely used in packaging applications for real-time freshness detection without multi-faceted analysis devices concerning consumer demand for safe food. In the last few years, intelligent indicators have been developed to replace synthetic chemical dyes using various natural pigments such as anthocyanins, curcumin, alizarin, and betalain.

10.5.1.3 *Chemical Treatment*

The chemical modification of a galactomannan film with glutaraldehyde could be successfully performed (56). Galactomannans are polysaccharides consisting of a mannose backbone with galactose side groups. Galactomannan is shown in Figure 10.2.

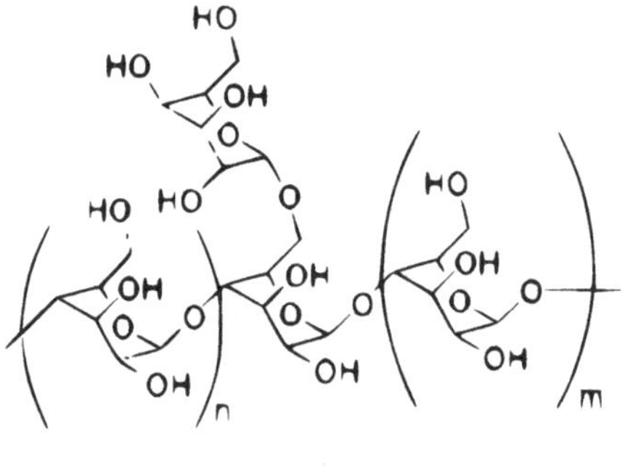

Figure 10.2 Galactomannan.

The results indicated that the UV irradiation with increasing dosage of photoinitiator could gradually consume the free hydroxyl groups in galactomannan to form the crosslinked networks, reducing the hydrophilic property and water vapor permeability of the modified film.

The tensile strength of the obtained film with 20% sodium benzoate addition was 2.8 times higher than a control sample. In comparison, aldol reaction between hydroxyl groups in galactomannan and glutaraldehyde resulted in a homogeneous surface of films, and the introduction of glutaraldehyde significantly increased the hydrophilicity of the modified films. After a chemical treatment, the value of elongation at break was remarkably enhanced by nearly 15 times, corresponding to the great flexible property against stretching (56).

10.5.1.4 UV Irradiation

The treatment of the surface of fresh fruits and vegetables with ultraviolet (UV) light has been presented in a review (57).

UV exposure, soil burial, and storage ability tests were carried out on biocomposite films with different concentrations of nanocrystal cellulose to study their biodegradability and morphological properties compared to neat whey protein films (58).

The addition of nanocrystal cellulose as a natural filler to the whey protein-based film can change a variety of properties such as strength, shelf life, biodegradability, and resistance to environmental conditions that are required in food packaging. The degradation films were assessed by UV exposure and soil burial tests.

In addition, the packaging ability was analyzed through storage tests under controlled conditions (58). Then, the changes in physical, chemical, and morphological properties were observed in the biocomposite film. From these results, it was concluded that the whey protein film with the addition of 3% nanocrystal cellulose would be a more favorable choice because of its better properties and suitability for the proposed biodegradable food packaging application (58).

The physical modification of a galactomannan film could be successfully performed with UV irradiation (56).

The results indicated that the UV irradiation with increasing dosage of photoinitiator could gradually consume the free hydroxyl groups in galactomannan to form the crosslinked networks, reducing the hydrophilic property and water vapor permeability of the modified film.

The tensile strength of the obtained film with 20% sodium benzoate addition was 2.8 times higher than a control sample. In comparison, aldol reaction between hydroxyl groups in galactomannan and glutaraldehyde resulted in a homogeneous surface of films, and the introduction of glutaraldehyde significantly increased the hydrophilicity of the modified films. After a chemical treatment, the value of elongation at break was remarkably enhanced by nearly 15 times, corresponding to the great flexible property against stretching (56).

10.5.1.5 Modified Atmosphere Packaging

The modified atmosphere packaging technology involves the alteration of the gas atmosphere inside a food package to preserve the food quality and extend its shelf life (59). This technology dates back to 1979 when Marks and Spencer introduced it commercially in the United Kingdom for the first time for processed meat products.

Modified atmosphere packaging provides an extension in food shelf life with minimum or no changes in the food attributes. This is achieved on the basis of modified atmosphere packaging principles, which are reduction in undesirable changes in food, control of microbial growth, and prevention of product contamination. The development of the modified atmosphere packaging technology and the contribution of polysaccharides to it were described (59).

The types of gases or gas mixtures used for different food product types have been detailed. Then, the effect of modified atmosphere packaging on the shelf life of different food types was elucidated, followed by a discussion of its effect on food nutritional quality and microorganisms.

The applications of various commonly used polysaccharides were emphasized. Alginate, cellulose, and chitosan were the polysaccharides mainly studied. Other less studied polysaccharides were gellan gum, pectin, carrageenan, and xanthan (59).

10.5.1.6 Active and Intelligent Packaging Films and Coatings

Seaweeds are used in active packaging, intelligent packaging, edible films, edible coatings, sachets, and in other utensils (60). To be used as food contact materials, seaweeds are combined with other polysaccharides, nanoparticles, essential oils, or plant extracts to improve their barrier thermal, mechanical, antioxidant, and antimicrobial properties. Seaweeds have been incorporated with other polymers or with seaweed polymers to form active packages such as alginate/sepiolite/myrtle berries extract, sodium alginate-β-cyclodextrin carvacrol, agar sodium alginate silicon dioxide nanoparticle, furcellaran/gelatin hydrolysate/rosemary extract.

With the incorporation of mulberry, anthocyanin, and the ability to change color at different pH sensitivities, seaweeds have also been used as intelligent packaging systems. Because of seaweeds non-lethal, non-toxic properties, as well as their film-forming abilities, these compositions were used as edible films and coatings (60).

10.5.2 Heat-Sealable Food Packing Films

Heat-sealable food packaging films, methods for the production thereof, and food packages that contain heat-sealable food packaging films have been described (61).

The heat-sealable food packaging films are made from a humidity-dependent permeable film having a moisture vapor transmission rate that increases with an increase in relative humidity. An outer coating is a coating material on at least one surface of the humidity-dependent permeable film.

The coating material is selected from a nanoclay dispersed in a poly(vinylidene chloride) polymer or a stretchable urethane polymer, a stretchable acrylic polymer, or a combination of stretchable urethane polymer and stretchable acrylic polymer. The coating material may further contain an anti-blocking agent. The heat-sealable food packaging film is biaxially oriented (61).

The humidity-dependent permeable film may be made from nylon. The biaxially oriented nylon film has a moisture vapor transmission rate (MVTR) of about 260 $g\,m^{-2}d^{-1}$ for 25 μm thickness at 23°C and 65% relative humidity, rising to over 400 $g\,m^{-2}d^{-1}$ and 90% relative humidity, but is not heat-sealable. Useful examples of nylons herein include homopolymers or copolymers selected from aliphatic poly(amide)s and aliphatic/aromatic polyamides having a molecular weight of from about 10 *kD* to about 100 *kD* (kilo Daltons).

General procedures useful for the preparation of poly(amide)s are well known to the art. Such procedures include the reaction products of diacids with diamines.

Useful diacids for making poly(amide)s are shown in Table 10.3 and in Figure 10.3. Suitable diamines for making poly(amide)s are shown in Table 10.4. Some diamines are shown in Figure 10.4. In addition, useful polyamide homopolymers are shown in Table 10.5.

Table 10.3 Diacids for poly(amide)s (61).

Aliphatic Dicarboxylic Acids	
Adipic acid	Sebacic acid
Glutaric acid	Pimelic acid
Suberic acid	Azelaic acid
Dodecanedioic acid	Octadecanedioic acid
Aromatic Dicarboxylic Acids	
Isophthalic acid	Terephthalic acid

Table 10.4 Diamines for poly(amide)s (61).

Diamine	Diamine
Trimethylene diamine	Tetramethylene diamine
Pentamethylene diamine	Hexamethylene diamine
Octamethylene diamine	Decamethylene diamine
Dodecamethylene diamine	Hexadecamethylene diamine
p-Phenylene diamine	4,4′-Diaminodiphenyl ether
4,4′-Diaminodiphenyl sulfone	4,4′-Diaminodiphenylmethane
2,2-Dimethylpentamethylene diamine	2,2,4-Trimethylhexamethylene diamine
2,4,4-Trimethylpentamethylene diamine	Diaminodicyclohexylmethane
Heptamethylene diamine	Nonamethylene diamine

Aliphatic poly(amide)s may be obtained from commercial sources or prepared in accordance with known preparatory techniques. Also, blends of two or more aliphatic/aromatic poly(amide)s can be used. Aliphatic/aromatic poly(amide)s can be prepared by known preparative techniques or can be obtained from commercial sources.

The humidity-dependent permeable film may be a poly(lactic acid) (PLA) film. PLA is a thermoplastic aliphatic polyester derived from renewable resources, such as corn starch, available in the United States, tapioca products, roots, chips or starch available mostly in Asia, or sugarcanes, available in the rest of world.

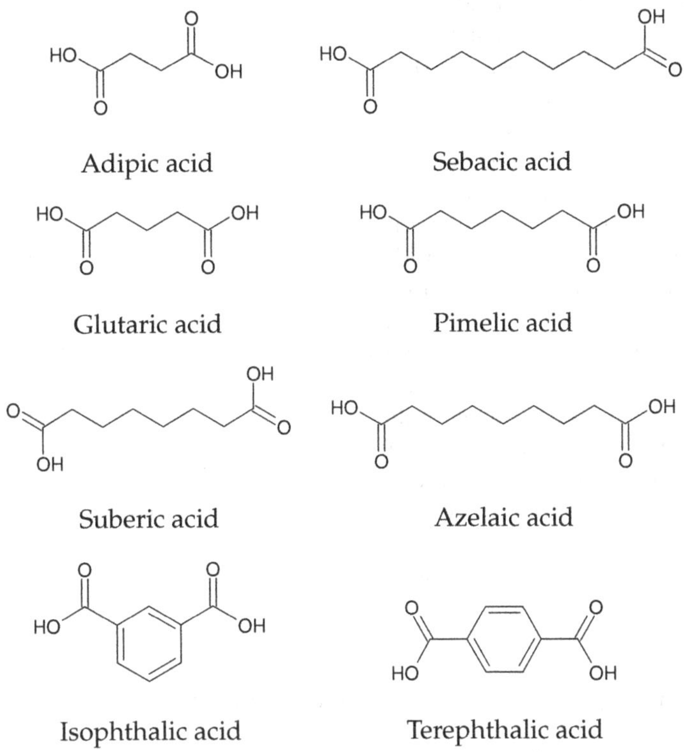

Figure 10.3 Diacids for Poly(amide)s (61).

Table 10.5 Poly(amide) polymers (61).

Homopolymer	Short Name
Poly(4-aminobutyric acid)	Nylon 4
Poly(6-aminohexanoic acid)	Nylon 6
Poly(7-aminoheptanoic acid)	Nylon 7
Poly(8-aminooctanoic acid)	Nylon 8
Poly(9-aminononanoic acid)	Nylon 9
Poly(10-aminodecanoic acid)	Nylon 10
Poly(11-aminoundecanoic acid)	Nylon 11
Poly(12-aminododecanoic acid)	Nylon 12
Poly(hexamethylene adipamide)	Nylon 6,6
Poly(hexamethylene sebacamide)	Nylon 6,10
Poly(heptamethylene pimelamide)	Nylon 7,7
Poly(octamethylene suberamide)	Nylon 8,8
Poly(hexamethylene azelamide)	Nylon 6,9
Poly(nonamethylene azelamide)	Nylon 9,9
Poly(decamethylene azelamide)	Nylon 10,9
Poly(tetramethylene diamine-*co*-oxalic acid)	Nylon 4,2
Polyamide of *n*-dodecanedioic acid and hexamethylene diamine	Nylon 6,12
Polyamide of dodecamethylene diamine and *n*-dodecanedioic acid	Nylon 12,12
Copolymer	**Short Name**
Caprolactam/hexamethylene adipamide copolymer	Nylon 6,6/6
Hexamethylene adipamide/caprolactam copolymer	Nylon 6/6,6
Trimethylene adipamide/hexamethylene azelaiamide copolymer	Nylon 6,2/6
Hexamethylene adipamide-hexamethylene-azelaiamide caprolactam copolymer	Nylon 6,6/6,9/6

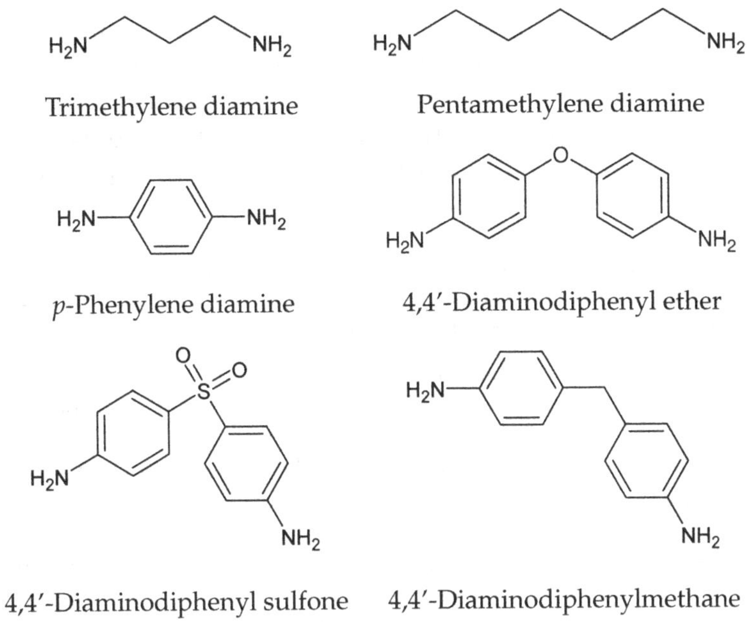

Figure 10.4 Diamines for poly(amide)s (61).

The humidity-dependent permeable film may be coated with the coating material to form a coated film. Conventional coating methods include, for example, gravure, reverse gravure, spray coating.

The coated film may optionally be dried in a heat setting step or still include water from the emulsion when subjected to stretching. The heat setting step may be performed at temperatures between about 65°C to about 120°C for about 2 *s* to about 30 *s* to substantially dry the coated film.

The coated film is then biaxially oriented, i.e., stretched in two perpendicular directions in a biaxial sheet stretching process. The coated film is stretched at an elevated temperature of about 190°C to about 220°C, enhancing adhesion of the coating material to the film, thereby avoiding the need for an intermediate adhesive primer layer (61).

10.5.3 Polymers for Food Packaging

The recent progress in polymer and polymer composite-enabled food packaging has been described (62).

Conventional synthetic undegradable polymers high-density poly(ethylene), low-density poly(ethylene), and linear low-density poly(ethylene), poly(styrene), poly(propylene), poly(ethylene terephthalate) are the main types of materials that are used in the modern food packaging industry. Some fields of application are shown in Table 10.6.

Table 10.6 Polymers for food packaging (63).

Polymer	Packaging Application
High-density poly(ethylene) (HDPE)	Custom packaging, grocery bags, water/milk/juice containers, cereal and snack liners
Low-density poly(ethylene) (LDPE)	Disposable cups, plates, and spoons; bread bags
Poly(styrene) (PS)	Take-away clamshells, meat trays, bottle caps, straws
Poly(propylene) (PP)	Drinking bottles for milk, food containers
Poly(ethylene terephthalate) (PET)	Water/juice/soft drink bottles, food jars, microwavable containers, plastic films
Poly(vinyl chloride) (PVC)	Plastic bags, frozen foods, stretch films, container lids

In addition to pure polymers, polymer composites (defined as polymers embedded with nanomaterials and organic or inorganic compounds) have also been intensively studied for advanced food packaging applications in recent years.

Nanomaterials have been composited with polymers not only as fillers to enhance their mechanical, thermal, and barrier properties, but also to bring many additional functions, e.g., antimicrobial, sensing, and detection (62).

Films and coatings are types of packaging materials used widely in the food industry for a variety of purposes (63). For example, they serve as protection against physical, chemical, as well as biological damage to food. Also, they are helpful in preventing the loss of aroma, flavor, antimicrobials, and antioxidants (64).

They prevent oxygen, carbon dioxide, and moisture loss, and are hence helpful in increasing the shelf life of food and improving the mechanical integrity and handling characteristics of food (65). The production and usage of petroleum-based polymer-generated plastic for manufacturing coatings and films have increased over recent decades. According to Derraik, plastic can be defined as a synthetic or semisynthetic organic polymer having good barrier properties against moisture, oil, and gases, with strength characteristics and good resistant properties against stress and corrosion (66). It is cheap, lightweight, and easy to carry and handle, making it highly convenient for daily use. Due to its abundant usage in various applications, the plastic industry in India is worth between 3000 and 4000 crores (67).

It is believed that half of all plastic products which are used in the food industry, such as cutlery, plastic bags, coatings, and packaging material, are designed to be thrown away. Moreover, the world production of plastics grew from 1.5 million metric tons in 1950 to 359 million metric tons in 2018 (68).

More than 35 million tons of waste derived from various plastic products are produced each year in the world and only 7% of these materials are recycled (69).

The usage and disposal of plastics has resulted in hazardous effects related to environmental as well as human health (70).

The current scenario of plastic pollution indicates that approximately 400,000 plastic bottles and about 700,000 plastic bags are disposed of worldwide every minute (69).

Since most polymers are non-biodegradable by nature, they remain in the environment, leading to severe environmental as well as health hazards. Most plastics and other petrochemical polymers are considered hard-to-degrade because of their corrosion-resistant qualities, and they can remain in the environment for decades or even centuries (71).

A study demonstrated that India generated almost 15,000 tons of plastic waste every day (72). This amount of plastic waste generation is a result of an increasing population, which reached 7.7 billion in 2019 and is expected to reach around 9.7 billion in 2050.

Biodegradable and edible films and coatings are an integral aspect of food packaging; they are formed from hydrocolloids such as proteins, polysaccharides, lipids and antimicrobial component. They consist of a continuous thin layer, between 0.050 *mm* and 0.250 *mm*, which is safe for consumption along with food (73, 74).

Films are mainly used to wrap food material, whereas coatings can be used directly on food products. The types of materials used for the production of these films are mostly renewable biological sources, mainly consisting of starch, cellulose, hemicellulose, protein, gelatin, lipids, fibers, etc. Examples of such materials include corn, sorghum, rice, fruit and vegetable waste, and agricultural waste such as wood chips or bagasse.

Waste produced by food industries, such as cane molasses from the sugar industry, barley spent grain from breweries, etc., could be a good source of these bio-based films as they are produced throughout the processing of food from agricultural land to food processing units to the consumer. They are easily available at a low cost and their usage would help in the waste utilization process (75),

Biodegradable and edible films that are produced from naturally available biomacromolecules have been seen as an alternative to synthetic polymers derived from petroleum in the development of food packaging. So, due to its rich composition and interesting bioactive properties, Spirulina has been considered a promising raw material for the production of multifunctional packaging materials.

However, unlike petroleum-based polymers, these materials do not offer good barrier and mechanical properties; therefore, some additives are also required to enhance these properties.

10.5.4 Polymer Membranes for Food Packaging

A method of packaging a food product using a polymer membrane has been developed (76). The polymer is a macromolecularly self-assembling material. The polymer is rendered into a film. The food product is packaged in an atmosphere with the polymer film wherein the film regulates the atmosphere in which the food product is packaged.

The method of packaging the food product includes fabricating packaging material through coinjection processes in the form of films, foils, sheets, containers or tubes among other packaging configurations (76).

Food products which may be subjected to the current method are fruits and vegetables, such as artichokes, apricots, banana, cranberry, eggplant, grapes, kiwifruit, leek, lettuce, mango, honeydew melons, mushrooms, nectarine, papaya, peach, pear, peas, bell pepper, pineapple, plum, radish, strawberries, and tomatoes among others, including mixtures thereof.

As useful materials, bis-amides, bis-urethanes and bis-urea units or their higher oligomers can be used. Molecularly self-assembling materials include molecularly self-assembling polyesteramides, copolyesteramide, copolyetheramide, copolyetherester-amide, copolyetherester-urethane, copolyether-urethane, copolyester-urethane, copolyester-urea, copolyetherester-urea and their mixtures. Preferred molecular self-assembly (MSA) includes copolyesteramide, copolyether-amide, copolyester-urethane, and copolyether-urethanes.

The preparation of a poly(esteramide) with 50 mol% of ethylene-*N,N'*-dihydroxyhexanamide material can be done in 4 steps as follows (76):

Step 1: Preparation of the diamide diol, ethylene-*N,N'*-dihydroxyhexanamide. A 10 *l* stainless steel reactor equipped with an agitator and a cooling water jacket is charged with ε-caprolactone (5.707 *kg*, 50 *mol*) and purged with nitrogen. Under rapid stirring, ethylene diamine (1.502 *kg* 25 *mol*) is added at once. After an induction period a slow exothermic reaction starts. The reactor temperature gradually rises to 90°C under maximum cooling applied. A white deposit forms and the reactor contents solidify, at which point stirring is stopped. The reactor contents are then cooled to 20°C and are then allowed to rest for 15 *h*. The reactor contents are then heated to 140°C (at which temperature the solidified reactor contents melt), and then heated further to 160°C under continued stirring for at least 2 *h*. The resulting liquid product is then discharged from the reactor into a collecting tray. A nuclear magnetic resonance study of the resulting product shows that the molar concentration of C2C in the product exceeds 80%. The procedure is repeated four more times, resulting in five product lots. The melting point of the product is determined to be 130—140°C (main melting point) by differential scanning calorimetry (DSC) (peak maximum). The solid material is granulated and used without further purification.

Step 2: Preparation of a MSA copolyesteramide with 50 mol%

amide content (PEA-C2C50%). Loading a reactor: A 100 *l* single-shaft Kneader-Devolatizer reactor equipped with a distillation column and a powerful vacuum pump system is nitrogen purged and heated to 80°C (thermostat oil). Dimethyl adipate (DMA) (38.324 *kg*) and granulated C2C monomer (31.724 *kg*, prepared as described above in Step 1) are fed into the kneader. The slurry is stirred at 50 revolutions per minute (rpm). 1,4-butane diol (1,4-BD, 18.436 kg*kg*) is added to the slurry at a temperature of about 60°C. The reactor temperature is further increased to 145°C to obtain a homogeneous solution.

Step 3: Distilling methanol and transesterification. Still under nitrogen atmosphere, titanium(IV)tetrabutoxide catalyst, 153 *g* in 1.380 *kg* 1,4-BD is injected at a temperature of 145°C in the reactor; methanol evolution starts. The temperature in reactor is slowly increased to 180°C in 1.75 *h* and is held for 45 additional minutes to complete the methanol distillation at ambient pressure. Methanol (12.664 *kg*) is collected.

Step 4: Distilling 1,4-butanediol and polycondensation to give PEA-C2C50%. The reactor dome temperature is increased to 130°C and the vacuum system activated stepwise to a reactor pressure of 7 *mbar* in 1 *h*. Temperature in the kneader/devolatizer reactor is kept at 180°C. Then the vacuum is increased to 0.7 *mbar* for 7 *h* while the temperature is increased to 190°C. The reactor is kept for 3 additional hours at 191°C and with vacuum ranging from 0.87 *mbar* to 0.75 *mbar*. At this point a sample of the reactor contents is taken (Sample Number 1); melt viscosities are 6575 *mPa s* at 180°C and 5300 *mPa s* at 190°C. The reaction is continued for another 1.5 *h* until a sample (Sample Number 2) shows final melt viscosities are 8400 *mPa s* at 180°C and 6575 *mPa s* at 190°C. Then the liquid kneader/devolatizer reactor contents are discharged at high temperatures of about 190°C into collecting trays, the resulting material is cooled to room temperature and ground. The weight of the final product PEA-C2C50% of Preparation 1 is 57.95 *kg* (87.8% yield). A sample (Sample Number 3) of the PEA-C2C50% of Preparation 1 has melt viscosities of 8625 *mPa s* at 180°C and 6725 *mPa s* at 190°C. Viscosities are determined using a Brookfied DV-II+ Vicosimeter with spindle number 28 at 20 *rpm* (revolutions per minute).

10.5.5 Natural Colorants

The effect of blended colorants of anthocyanin and shikonin on carboxymethyl cellulose/agar-based smart packaging film has been described (77).

10.5.5.1 Curcumin

Curcumin, one of the most well-known natural colorants, has received a lot of attention for its excellent functional properties and ability to change color with changes in pH. Curcumin is shown in Figure 10.5.

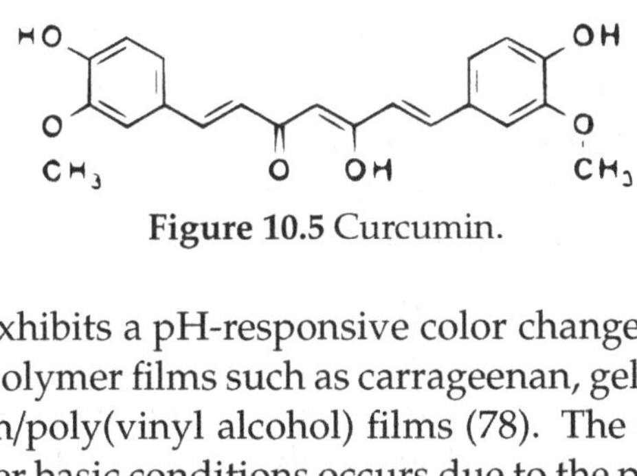

Figure 10.5 Curcumin.

Curcumin exhibits a pH-responsive color change when incorporated into biopolymer films such as carrageenan, gelatin, poly(lactic acid), and gum/poly(vinyl alcohol) films (78). The color change of curcumin under basic conditions occurs due to the phenoxide anion formed by the reaction of phenolic hydroxyl groups with hydroxyl ions (79).

Curcumin, the golden component of turmeric, a spice widely used in food since ancient times, is a cost-effective and abundant biomaterial with various biological properties such as antioxidant, antibacterial, antiviral, antitumor, and anti-inflammatory.

Today, active and intelligent packaging systems are at the forefront of cutting-edge technologies for the food supply chain (52, 53).

An active food packaging film is a preformed, thin-layer, free-standing film with functional properties that can be used in food packaging, driven by the consumer demand for high-quality, safe, natural foods.

10.6 Food Container

A food container can be a part of a plastic coffee or tea capsule. It is also suitable for receiving milk products or brewed products other than coffee or tea, such as for instant beverages or instant soups (80).

Plastic food containers are usually not impermeable or gas-tight, and especially not sealed against the ingress of oxygen, which can interact with the food contained in the food container.

When a food container is used for oxygen-sensitive foods, it is thus preferably provided with a gas-tight, water-tight, flavor-tight, and food-safe coating and/or barrier layer if one wishes to reduce to a minimum or even prevent losses in quality due to the effects of oxygen on the foods contained within it. Moreover, a barrier layer of this type provides a flavor seal that protects the foods contained in the food container from losing their flavor, even after a longer period of time. A water vapor barrier can be additionally provided when the food container is used as a receiving unit for products to be brewed (brewing products) so that water-soluble substances in the food container or in its cover film are not melted or even dissolved during the brewing process (80).

A plasma evaporation method such as this functions with silicon oxides, which are evaporated in a high vacuum by means of an electric arc, which is generated electrically by high voltage, wherein the silicon oxide vapor subsequently settles on the objects located in the vacuum chamber, including the coffee capsules.

This method is used for the coating of food packaging unit containers, wherein x should represent a number <2, preferably <0.9, as a further coating method for achieving oxygen impermeability (80).

10.7 Future Trends in Summary

This chapter dealt with polymers for food applications. Polymers are mainly used for food packaging. However, there are also polymer types that are eatable and can be used directly in food.

References

1. C.C. Villa, L.T. Sánchez, G.A. Valencia, S. Ahmed, and T.J. Gutiérrez, Molecularly imprinted polymers for food applications: A review, *Trends in Food Science & Technology* 642–669 (2021).
2. Y. Zhou, H. Liu, J. Li, Z. Sun, T. Cai, X. Wang, S. Zhao, and B. Gong, Restricted access magnetic imprinted microspheres for directly selective extraction of tetracycline veterinary drugs from complex samples, *Journal of Chromatography A* 460684 (2020).
3. Y.-K. Lv, C.-X. Zhao, P. Li, Y.-D. He, Z.-R. Yang, and H.-W. Sun, Preparation of doxycycline-imprinted magnetic microspheres by inverse-emulsion suspension polymerization for magnetic dispersion extraction of tetracyclines from milk samples, *Journal of Separation Science* 2656–2663 (2013).
4. Y. Zhang, M. Pei, B. Zhang, Y. He, and Y. Zhong, Changes of antibiotic resistance genes and bacterial communities in the advanced biological wastewater treatment system under low selective pressure of tetracycline, *Water Research* 117834 (2021).
5. A.Y. Udalova, S. Dmitrienko, and V. Apyari, Methods for the separation, preconcentration, and determination of tetracycline antibiotics, *Journal of Analytical Chemistry* 661–676 (2015).
6. N. Li, L. Zhou, X. Jin, G. Owens, and Z. Chen, Simultaneous removal of tetracycline and oxytetracycline antibiotics from wastewater using a ZIF-8 metal organic-framework, *Journal of Hazardous Materials* 563–572 (2019).
7. S. Han, Q. Leng, F. Teng, Y. Ding, and A. Yao, Preparation of mesh covalent organic framework Tppa-2-based adsorption enhanced magnetic molecularly imprinted composite for selective extraction of tetracycline residues from animal-derived foods, *Food Chemistry* 132601 (2022).
8. A. Wu, *Food Safety & Mycotoxins*, Springer, (2019).
9. A. Alshannaq and J.-H. Yu, Occurrence, toxicity, and analysis of major mycotoxins in food, *International Journal of Environmental Research and Public Health* 632 (2017).
10. M. Eskola, G. Kos, C.T. Elliott, J. Hajšlová, S. Mayar, and R. Krska, Worldwide contamination of food-crops with mycotoxins: Validity of the widely cited FAO estimateof 25%, *Critical Reviews in Food Science and Nutrition* 2773–2789 (2020).
11. Y. Hua, Y. Ahmadi, C. Sonne, and K.-H. Kim, Progress and challenges in sensing of mycotoxins using molecularly imprinted polymers, *Environmental Pollution* 119218 (2022).
12. B. Teli, J. Purohit, M.M. Rashid, A.A.K. Jailani, and A. Chattopadhyay, Omics insight on Fusarium head blight of wheat for translational research perspective, *Current Genomics* 411–428 (2020).

13. A. Speltini, A. Scalabrini, F. Maraschi, M. Sturini, and A. Profumo, Newest applications of molecularly imprinted polymers for extraction of contaminants from environmental and food matrices: A review, *Analytica Chimica Acta* 1–26 (2017).
14. G.A. Valencia, E.N. Zare, P. Makvandi, and T.J. Gutiérrez, Self-assembled carbohydrate polymers for food applications: A review, *Comprehensive Reviews in Food Science and Food Safety* 2009–2024 (2019).
15. J. Penelle, MIP/QCM sensors for high sensitivity-fast sensing of small molecules in solution, US Patent 6 890 486, assigned to University of Massachusetts UMass, May 10, 2005.
16. A.B.F. Martinson, J.A. Libera, J.W. Elam, and S.C. Riha, Micro-balance sensor integrated with atomic layer deposition chamber, US Patent 9 856 563, assigned to UChicago Argonne LLC, January 02, 2018.
17. R.C. Ebersole, R.P. Foss, and M.D. Ward, Analyte detection by means of an analyte-responsive polymer, US Patent 5 756 279, assigned to E.I. Du Pont de Nemours and Co, May 26, 1998.
18. I. Karube and H.C.S.I. Muramatsu, Analytical device and method for analysis of biochemicals, microbes and cells, assigned to Seiko Instruments Inc.
19. D. Mueller-Schulte and H. Laurs, Means and process for manufacturing biosensors on the basis of an oscillating quartz crystal, and use thereof for determining antigens and affinity ligands, DE Patent 3 733 986, assigned to Mueller Schulte Detlef Dr, April 20, 1989.
20. Wikipedia contributors, Sauerbrey equation — Wikipedia, the free encyclopedia, https://en.wikipedia.org/wiki/Sauerbrey_equation, (2023). [Online; accessed 03-August-2023].
21. R.P. Foss and D.R. Fruge, Wash-off film containing synthetic amphoteric polymers, US Patent 4 735 887, assigned to E.I. Du Pont de Nemours and Co, April 05, 1988.
22. R.C. Ebersole, R.P. Foss, and M.D. Ward, Piezoelectric cell growth biosensing method and system, WO Patent 1 991 001 381, assigned to E.I. Du Pont de Nemours And Co, February 07, 1991.
23. T. Tanaka and Y. Hirokawa, Reversible, discontinuous volume changes of ionized isopropylacrylamide cells, US Patent 4 732 930, assigned to Massachusetts Institute of Technology, March 22, 1988.
24. D.W. Schroll, *Flammability and Sensitivity of Materials in Oxygen-enriched Atmospheres: Third Volume*, Vol. 3, ASTM International, (1988).
25. R.P. Foss, Acrylic amphoteric polymers, US Patent 4 749 762, assigned to E.I. Du Pont de Nemours and Co., June 07, 1988.
26. F. Fauzi, A. Rianjanu, I. Santoso, and K. Triyana, Gas and humidity sensing with quartz crystal microbalance (QCM) coated with graphene-based materials–A mini review, *Sensors and Actuators A: Physical* 112837 (2021).

27. A. Rianjanu, E. Nurfani, M.F. Arif, K. Triyana, and H.S. Wasisto, Stability evaluation of quartz crystal microbalances coated with polyvinyl acetate nanofibrous mats as butanol vapor sensors, *Materials Today Communications* 101770 (2021).
28. L. Wang, Metal-organic frameworks for QCM-based gas sensors: A review, *Sensors and Actuators A: Physical* 111984 (2020).
29. K. Liu and C. Zhang, Volatile organic compounds gas sensor based on quartz crystal microbalance for fruit freshness detection: A review, *Food Chemistry* 127615 (2021).
30. P. Hampitak, D. Melendrez, M. Iliut, M. Fresquet, N. Parsons, B. Spencer, T.A. Jowitt, and A. Vijayaraghavan, Protein interactions and conformations on graphene-based materials mapped using a quartz-crystal microbalance with dissipation monitoring (QCM-D), *Carbon* 317–327 (2020).
31. D.B. Nugroho, A. Rianjanu, K. Triyana, A. Kusumaatmaja, and R. Roto, Quartz crystal microbalance-coated cellulose acetate nanofibers overlaid with chitosan for detection of acetic anhydride vapor, *Results in Physics* 102680 (2019).
32. A. Rianjanu, D.B. Nugroho, A. Kusumaatmaja, R. Roto, and K. Triyana, A study of quartz crystal microbalance modified with polyvinyl acetate nanofiber to differentiate short-chain alcohol isomers, *Sensing and Bio-Sensing Research* 100294 (2019).
33. A. Rianjanu, K. Triyana, D.B. Nugroho, A. Kusumaatmaja, and R. Roto, Electrospun polyvinyl acetate nanofiber modified quartz crystal microbalance for detection of primary alcohol vapor, *Sensors and Actuators A: Physical* 111742 (2020).
34. Z. Wu, X. Chen, S. Zhu, Z. Zhou, Y. Yao, W. Quan, and B. Liu, Enhanced sensitivity of ammonia sensor using graphene/polyaniline nanocomposite, *Sensors and Actuators B: Chemical* 485–493 (2013).
35. X. Li, X. Chen, Y. Yao, N. Li, and X. Chen, High-stability quartz crystal microbalance ammonia sensor utilizing graphene oxide isolation layer, *Sensors and Actuators B: Chemical* 183–188 (2014).
36. P. Larki, Y.M. Sabri, K.M. Kabir, A. Nafady, A.E. Kandjani, and S.K. Bhargava, Silver/gold core/shell nanowire monolayer on a QCM microsensor for enhanced mercury detection, *RSC Advances* 92303–92311 (2015).
37. L. Wang, X. Cha, Y. Wu, J. Xu, Z. Cheng, Q. Xiang, and J. Xu, Superhydrophobic polymerized *n*-octadecylsilane surface for BTEX sensing and stable toluene/water selective detection based on QCM sensor, *ACS Omega* 2437–2443 (2018).
38. M. Yang, J. He, X. Hu, C. Yan, and Z. Cheng, CuO nanostructures as quartz crystal microbalance sensing layers for detection of trace hydrogen cyanide gas, *Environmental Science & Technology* 6088–6094 (2011).

39. C. Zheng, L. Zhu, J. Wang, et al., Tin sulfides heterostructure modified quartz crystal microbalance sensors with high sensitivity for hazardous trimethylamine gas, *Sensors and Actuators B: Chemical* 132520 (2022).
40. O. Alev, O. Özdemir, E. Goldenberg, L.C. Arslan, S. Büyükköse, and Z.Z. Öztürk, Ws2 thin film based quartz crystal microbalance gas sensor for dimethyl methylphosphonate detection at room temperature, *Thin Solid Films* 139097 (2022).
41. P. Qi, Z. Xu, T. Zhang, T. Fei, and R. Wang, Chitosan wrapped multiwalled carbon nanotubes as quartz crystal microbalance sensing material for humidity detection, *Journal of Colloid and Interface Science* 284–292 (2020).
42. S. Farooq, H. Wu, J. Nie, S. Ahmad, I. Muhammad, M. Zeeshan, R. Khan, and M. Asim, Application, advancement and green aspects of magnetic molecularly imprinted polymers in pesticide residue detection, *Science of the Total Environment* 150293 (2022).
43. Q. Chen, M. Meng, W. Li, Y. Xiong, Y. Fang, and Q. Lin, Emerging biosensors to detect aflatoxin M1 in milk and dairy products, *Food Chemistry* 133848 (2022).
44. K. Bao, X. Liu, Q. Xu, B. Su, Z. Liu, H. Cao, and Q. Chen, Nanobody multimerization strategy to enhance the sensitivity of competitive ELISA for detection of ochratoxin A in coffee samples, *Food Control* 108167 (2021).
45. J. Xiong, L. Xiong, H. Zhou, Y. Liu, and L. Wu, Occurrence of aflatoxin B1 in dairy cow feedstuff and aflatoxin M1 in UHT and pasteurized milk in central China, *Food Control* 386–390 (2018).
46. Y. Xiong, K. Pei, Y. Wu, H. Duan, W. Lai, and Y. Xiong, Plasmonic ELISA based on enzyme-assisted etching of Au nanorods for the highly sensitive detection of aflatoxin B1 in corn samples, *Sensors and Actuators B: Chemical* 320–327 (2018).
47. B. Li, Y. Wang, H. Wei, and S. Dong, Amplified electrochemical aptasensor taking AuNPs based sandwich sensing platform as a model, *Biosensors and Bioelectronics* 965–970 (2008).
48. Y.S. Kim, J.H. Kim, I.A. Kim, S.J. Lee, J. Jurng, and M.B. Gu, A novel colorimetric aptasensor using gold nanoparticle for a highly sensitive and specific detection of oxytetracycline, *Biosensors and Bioelectronics* 1644–1649 (2010).
49. A.J. Barbosa, A.R. Oliveira, and A.C. Roque, Protein-and peptide-based biosensors in artificial olfaction, *Trends in Biotechnology* 1244–1258 (2018).
50. A. Karimzadeh, M. Hasanzadeh, N. Shadjou, and M. de la Guardia, Peptide based biosensors, *TrAC Trends in Analytical Chemistry* 1–20 (2018).

51. S. Roy, R. Priyadarshi, P. Ezati, and J.-W. Rhim, Curcumin and its uses in active and smart food packaging applications - A comprehensive review, *Food Chemistry* 131885 (2022).
52. S. Mohammadalinejhad, H. Almasi, and M. Moradi, Immobilization of Echium amoenum anthocyanins into bacterial cellulose film: A novel colorimetric pH indicator for freshness/spoilage monitoring of shrimp, *Food Control* 107169 (2020).
53. M.S. Firouz, K. Mohi-Alden, and M. Omid, A critical review on intelligent and active packaging in the food industry: Research and development, *Food Research International* 110113 (2021).
54. S. Davoodi, S.M. Davachi, A. Ghorbani Golkhajeh, A.S. Shekarabi, and A. Abbaspourrad, Development and characterization of Salvia macrosiphon/Chitosan edible films, *ACS Sustainable Chemistry & Engineering* 1487–1496 (2020).
55. R. Priyadarshi, S.-M. Kim, and J.-W. Rhim, Pectin/pullulan blend films for food packaging: Effect of blending ratio, *Food Chemistry* 129022 (2021).
56. N. Zhao, Y. Chai, T. Wang, K. Wang, J. Jiang, and H.-y. Yang, Preparation and physical/chemical modification of galactomannan film for food packaging, *International Journal of Biological Macromolecules* 1060–1067 (2019).
57. M. Turtoi, Ultraviolet light treatment of fresh fruits and vegetables surface: A review, *Journal of Agroalimentary Processes and Technologies* 325–337 (2013).
58. F. Fitriani, S. Aprilia, N. Arahman, M.R. Bilad, F. Djuned, U. Fathanah, H. Meilina, and A. Amin, Studies of exposure UV light, soil Burial, and storage ability effect on characteristics of biocomposite films, *Materials Today: Proceedings* (2023).
59. R. Priyadarshi, F. Deeba, Y.S. Negi, et al., Modified atmosphere packaging development in *Processing and Development of Polysaccharide-Based Biopolymers for Packaging Applications* 261–280. Elsevier, (2020).
60. K.Y. Perera, S. Sharma, D. Pradhan, A.K. Jaiswal, and S. Jaiswal, Seaweed polysaccharide in food contact materials (active packaging, intelligent packaging, edible films, and coatings), *Foods* 2088 (2021).
61. Y.-P.R. Ting, S.J. Porter, K. Guhse, and J.D. Moulton, Heat sealable food packing films, methods for the production thereof, and food packages comprising heat sealable food packaging films, US Patent 10 435 225, assigned to AdvanSix Resins and Chemicals LLC, October 08, 2019.
62. M. Zhang, G.M. Biesold, W. Choi, J. Yu, Y. Deng, C. Silvestre, and Z. Lin, Recent advances in polymers and polymer composites for food packaging, *Materials Today* 134–161 (2022).

63. V. Gupta, D. Biswas, and S. Roy, A comprehensive review of biodegradable polymer-based films and coatings and their food packaging applications, *Materials* 5899 (2022).
64. E. Díaz-Montes and R. Castro-Muñoz, Edible films and coatings as food-quality preservers: An overview, *Foods* 249 (2021).
65. F. Versino, O.V. Lopez, M.A. Garcia, and N.E. Zaritzky, Starch-based films and food coatings: An overview, *Starch-Stärke* 1026–1037 (2016).
66. J.G. Derraik, The pollution of the marine environment by plastic debris: A review, *Marine Pollution Bulletin* 842–852 (2002).
67. V. Kumbhar and P. Masali, Biodegradable cutlery using Moringapod husk: An alternative to conventional plastic cutlery, *Int. J. Innov. Sci. Res. Technol* 900–903 (2020).
68. M. Gardside, Global plastic production from 1950 to 2018 (in million metric tons), (2020).
69. M. Sabbah and R. Porta, Plastic pollution and the challenge of bioplastics, *J. Appl. Biotechnol. Bioeng* 00033 (2017).
70. A.A. de Souza Machado, W. Kloas, C. Zarfl, S. Hempel, and M.C. Rillig, Microplastics as an emerging threat to terrestrial ecosystems, *Global Change Biology* 1405–1416 (2018).
71. W.C. Li, H.F. Tse, and L. Fok, Plastic waste in the marine environment: A review of sources, occurrence and effects, *Science of the Total Environment* 333–349 (2016).
72. H.N. Patil and P. Sinhal, A study on edible cutlery: An alternative for conventional ones, *Atithya: A Journal of Hospitality* 45–51 (2018).
73. X. Luo, J. Zhao, M. Li, X. Zhao, X. Wei, Z. Luo, W. Gu, D. Du, Y. Lin, and C. Zhu, Single-atom materials for food safety, *Materials Today* (2023).
74. R. Schoonjans, J. Castenmiller, Q. Chaudhry, F. Cubadda, T. Daskaleros, R. Franz, D. Gott, J. Mast, A. Mortensen, A.G. Oomen, et al., Regulatory safety assessment of nanoparticles for the food chain in Europe, *Trends in Food Science & Technology* 98–111 (2023).
75. M.M. Nakamoto, M. Assis, J.G. de Oliveira Filho, and A.R.C. Braga, Spirulina application in food packaging: Gaps of knowledge and future trends, *Trends in Food Science & Technology* (2023).
76. S.T. Matteucci, L.C. Lopez, S.D. Feist, P.N. Nickias, and S.N. Khot, Self assembling polymer membranes in food packaging application, US Patent 9 193 487, assigned to Dow Global Technologies LLC, November 24, 2015.
77. S. Roy, H.-J. Kim, and J.-W. Rhim, Effect of blended colorants of anthocyanin and shikonin on carboxymethyl cellulose/agar-based smart packaging film, *International Journal of Biological Macromolecules* 305–315 (2021).
78. N. Aliabbasi, M. Fathi, and Z. Emam-Djomeh, Curcumin: A promising bioactive agent for application in food packaging systems, *Journal of Environmental Chemical Engineering* 105520 (2021).

79. J. Liu, H. Wang, P. Wang, M. Guo, S. Jiang, X. Li, and S. Jiang, Films based on κ-carrageenan incorporated with curcumin for freshness monitoring, *Food Hydrocolloids* 134–142 (2018).
80. N. Kuhl and M. Wassmer, Oxygen-tight food container, US Patent 20 230 075 620, assigned to Individual, March 09, 2023.

11

Additive Classes

11.1 Compatibilizers

Additives are a special class of compatibilizers (1).

The compatibilization of immiscible polymers has proved to be a challenge since the genesis of polymer science and technology and has re-emerged in recent years as a theme of key importance to the plastic waste problem (2).

Chemically dissimilar polymers are rarely miscible over the full composition range due to an entropy of mixing per unit volume that scales as 1/N, where N is the degree of polymerization of a chain. Miscible polymer alloys of high molecular weight thus require either a remarkable structural similarity (e.g., isotopes or isomers) or specific attractions between dissimilar segments (e.g., via hydrogen bonds) to render the enthalpy of mixing vanishingly small or negative.

Truly compatible polymer alloys are rare, and only a few are exploited commercially (e.g., poly(styrene)/poly(*p*-phenylene oxide)) (3).

More typically, polymer blends exhibit macrophase separation and have poor mechanical properties due to narrow interfaces with low chain entanglement between coexisting phases/domains. Other properties, including optical clarity and electronic/ionic conductivity, can be similarly limited by ca. 1 μm domains separated by amorphous interfaces. The physical properties of immiscible blends are also unpredictable with melt or solution processing time and history due to a propensity for microstructural evolution (4).

A variety of strategies have been developed to compatibilize polymer blends, including the addition of a block or graft copolymer to lower the interfacial tension between the coexisting domains, broaden and strengthen interfaces, and stabilize multidomain morphologies against coarsening. In many cases, the block or graft copolymer is created in-situ by reactive blending wherein the copolymer is produced by a chemical reaction between functional groups on the dissimilar polymers. Reactive blending is typically conducted in the melt state within an extruder where mixing and reaction simultaneously occur (5). The copolymer that is either blended or reactively formed in an extruder most commonly has dissimilar blocks/grafts joined by permanent covalent bonds or reversible hydrogen bonds (6).

However, there is a smaller body of literature on copolymers produced by ionic bonds formed by combining acidic units on one polymer (e.g., a sulfonic or carboxylic acid) with basic units on a second polymer (e.g., an amine, pyridine, or imidazole) (7, 8).

The proton exchange that occurs between a pair of such units leaves charges of opposite sign on the two polymers, which in a low dielectric environment creates a strong ionic bond. Importantly, no counterions are generated by such a reaction, so the electrostatic interactions are strong and weakly screened. Alternatively, salts of the acid and base functionalized polymers can be used to produce ionic bonding (4), leaving residual small molecule counterions in that case. There is a 30+ year body of literature on ionic bonding in polymers, including well-defined systems with terminal functional groups (9). So, this remains a powerful tool for compatibilizing broad combinations of polymers (2).

11.2 Contaminants

Contaminants may include hazardous materials, legacy additives that have been phased out or are banned for specific use (10), odor-bearing compounds, chemically active substances which, for example, catalyze polymer degradation, or ingredients imparting undesired color.

11.2.1 *Weathering of Plastic Materials*

Increases in plastic-related pollution and their weathering can be a serious threat to environmental sustainability and human health, especially during the COVID-19 pandemic (11). Planetary risks of plastic waste disposed from diverse sources are exacerbated by the weathering-driven alterations in their physical-chemical attributes and the presence of hazardous pollutants mediated through adsorption.

Plastic polymers act as vectors of toxic chemical contaminants and pathogenic microbes through sorption onto the plastic-microbe/biofilm-environment interface.

The effects of weathering-driven alterations on the plastisphere have been addressed in relation to environmental contaminants. Contaminants are endocrine-disrupting chemicals, polycyclic aromatic hydrocarbons, polychlorinated biphenyl compounds, pharmaceuticals and personal care products.

The detailed evaluation on the role of complex associations between weathering and physicochemical properties of plastics should help to gain better knowledge with respect to the transport, behavior, fate, and toxicological chemistry of plastics along with the proper tactics for their sustainable remediation (11).

11.2.2 *Contaminants in Wastewater*

The increased amounts of contaminants in waters challenges the scientific community and water treatment stakeholders to design remediation techniques that are simple, practical, inexpensive, effective, and environmentally friendly (12).

Emerging contaminants include antibiotics, hormones, illicit drugs, endocrine disruptors, cosmetics, personal care products, pesticides, surfactants, industrial products, microplastics, nanoparticles, and nanomaterials. The removal of those contaminants is not easy because classical wastewater treatment systems are not designed to handle emerging contaminants, and contaminants often occur as traces in complex organo-mineral mixtures.

Advanced treatments for the removal of emerging contaminants in wastewater were also described. Adsorption-oriented processes using non-conventional adsorbents were also described, such as cyclodextrin polymers, metal organic frameworks, molecularly imprinted polymers, chitosan, and nanocellulose.

Liquid-solid adsorption using commercial activated carbons is a key method for removing contaminants from wastewater or drinking water sources. Indeed, this industrial technique of contaminant removal can produce high quality water, while also being a process that is both technologically simple and economically feasible.

Commercial activated carbons in granular or powder form are effective adsorbents for treating a wide range of contaminants, including pesticides and pharmaceuticals, for adsorbing organic matter to reduce chemical oxygen demand and biochemical oxygen demand, for decoloring water, or for treating substances that cause a specific taste or odor (12). However, even though the high adsorption capacity of active carbons is widely recognized, there are some drawbacks, namely their rapid saturation and regeneration. This regeneration step of saturated carbon is costly, not straightforward, and leads to a loss of the adsorbent. However, even though active carbons are highly absorbent, they have several drawbacks.

Active carbons are quite expensive, the higher the quality, the greater the cost. The different qualities of carbon are strongly related to the raw material used, as well as dependent on the carbonization conditions and the way in which activation is carried out, either physically or chemically. There are also two shortcomings, the rapid saturation and the need of regeneration of saturated carbon, which is also expensive, not straightforward, and results in adsorbent loss.

Therefore it is interesting to develop new materials capable of having the same performance as carbon without its disadvantages. Numerous studies have been published to meet this challenge. The research is divided into two directions: the use of natural, cheap, and abundant non-conventional materials in raw or modified form and the development of new synthetic materials.

The recently proposed materials include cyclodextrin beads, metal-organic frameworks, molecularly imprinted polymers, chitosan-based materials, and nanocellulose.

Also, biological-based technologies for the degradation and removal of emerging contaminants have been described. Advanced oxidation processes are the most promising strategies because of their simplicity and efficiency (12).

11.2.3 *Removal of Pharmaceuticals Using Cyclodextrin*

Much work has been done on the use of cyclodextrin polymers for the manufacture of complex emerging substances for environmental applications.

These polymers have become very popular for the tracking and removal of trace emerging substances. Conventional water treatment technologies are not very effective in reducing the concentration of these pollutants to a desirable level. So, complementary methods to conventional treatments were proposed, such as the use of cyclodextrin bead polymers.

Cyclodextrins are carbohydrates produced from starch and characterized by cyclic structure with hydrophilic surface and moderately hydrophobic inner cavity; cyclodextrins are able to encapsulate compounds in their cavity.

Within the cavity of cyclodextrins, the water molecules are readily replaced by less hydrophilic guest compounds such as low water-soluble drugs. In this way, inclusion complexes or clathrates are formed without creating or disrupting covalent bonds.

The host cyclodextrin and the guest are reversibly connected by hydrophobic interactions and van der Walls forces. Therefore, the complex can dissociate when the conditions change, e.g., upon dilution, heating or in the presence of other competing guests.

The stability of the inclusion complexes can be enhanced by hydrogen bonds between the hydroxyl groups on the outer surface of cyclodextrin and the hydrophilic part of the guest protruding from the cavity. The most important consequences of the inclusion complex formation, such as enhancing the solubility of poorly soluble substances, protecting the included guest molecules against the environmental effects such as oxidation, hydrolysis, decomposition on heat and light, enzymatic degradation, and improving the taste, are utilized broadly by the pharmaceutical industry.

The number of active ingredients marketed in the form of cyclodextrin complexes was above 100 at the end of 2021 and is continuously increasing. Among the three natural cyclodextrins, α-cyclodextrin, β-cyclodextrin and γ-cyclodextrin consisting of 6, 7 and 8 glucopyranose units, respectively, β-cyclodextrin possesses a cavity size with an internal diameter of 0.60 to 0.65 *nm* and a depth of 0.78 *nm* especially suitable for inclusion of a wide range of drug molecules.

While the unmodified β-cyclodextrin is used mostly as excipient in pharmaceutical formulations for oral administration, its highly soluble hydroxypropylated and sulfobutylated derivatives have also been approved for parenteral applications.

While the pharmaceutical industry aims to improve the solubility and consequently the bioavailability of drugs via complexation, the objective in wastewater purification is the opposite: To remove the dissolved non-biodegradable pharmaceuticals by adsorption/-sorption/biosorption, both terms being used in the literature. As natural cyclodextrins are water-solubles, they must be transformed into water-insoluble sorbents/biosorbents, using three main synthesis routes (12):

1. Crosslinking with proper bifunctional or polyfunctional reagents, such as diepoxy compounds, diisocyanates, dicarboxylic acids or polycarboxylic acids, and fluoro terphthalonitriles,
2. Copolymerization of polymerizable cyclodextrin derivatives with acrylic or vinyl monomers,
3. Grafting on a macromolecular support such as chitosan, cellulose, alginate, or silica.

All these gels are preferably prepared in the form of tiny beads by emulsion/suspension polymerization (12).

The spherical shape of the particles has the advantage of high surface area and good accessibility of the cyclodextrin cavities in addition to the technological advantages in scaled-up production. Owing to the several hydroxyl groups on cyclodextrins, these gels are highly hydrophilic and swell in water. The degree of swelling depends on the conditions of preparation, e.g., type, concentration and ratio of reactants, temperature, and reaction time.

The higher the degree of swelling the higher is the accessibility of the inner cyclodextrin cavities, but the smaller the mechanical stability against compression. The binding capacity of epichlorohydrin-crosslinked β-cyclodextrin bead polymer is higher in the case of lower degree of crosslinking and reduced temperature.

For ionic or ionizable pharmaceuticals, cyclodextrin polymers modified by ionic groups can be more efficient biosorbents due to the contribution of electrostatic forces to hydrophobic interactions. Depending on the conditions, the binding mechanism can be complicated, including inclusion and association complex formation, electrostatic interactions, ion exchange, and chelation (12).

11.3 Legacy Additives

Many plastic products can only achieve their specific characteristics if mixed with additives like flame retardants, stabilizers, and plasticizers. Some of the formerly used plastic additives, however, are in the meantime evaluated as substances of very high concern (SVHC) or even persistent organic pollutants (POP) and are referred to as *legacy additives* (10).

Therefore, the improper disposal of legacy plastic waste as well as the recycling and reuse of such can lead to continuous circulation of harmful additives into the environment, threatening plant and animal life, and human health.

The environmental threats posed by hazardous additives have been addressed by international regulations like the Registration, Evaluation, and Authorization of Chemicals (REACH) regulation and the Stockholm Convention. They set thresholds for hazardous chemicals including some legacy additives, which regulate trading and waste management, and aim at a fast phase-out of hazardous compounds. At the same time, governmental, non-governmental and industrial players support a circular economy of goods including plastics, which resulted and will further result in increasing recycling quotas for waste plastics. Hazardous additives are shown in Table 11.1 and in Figure 11.1.

11.4 Chain Extenders

The urgent need to enhance the recycling abilities of high volume non-degradable plastics, such as poly(ethylene terephthalate), poly(carbonate) and Nylon, has driven the development of additives called chain extenders (13).

Chain extension with designed multifunctional polymers is introduced to enhance the properties of sustainable (bio)polymers, such as poly(lactic acid) and poly(β-hydroxyalkanoate), enabling them

Table 11.1 Hazardous additives (10).

Additive	Usage
Hexabromocyclododecan	Flame retardant
Tetrabromodiphenyl ether	Flame retardant
Pentabromodiphenyl ether	Flame retardant
Hexabromodiphenyl ether	Flame retardant
Heptabromodiphenyl ether	Flame retardant
Decabromodiphenyl ether	Flame retardant
Polybrominated biphenyl	Flame retardant, plasticizer
Diethylhexyl phthalate	Plasticizer
Benzylbutyl phthalate	Plasticizer
Dibutyl phthalate	Plasticizer
Diisobutyl phthalate	Plasticizer
Lead compounds	Stabilizer
Cadmium compounds	Stabilizer

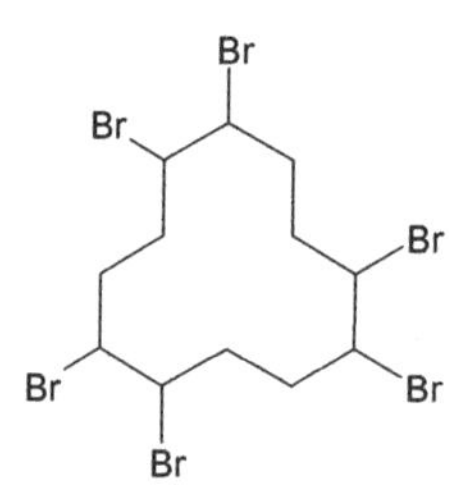

Hexabromocyclododecan

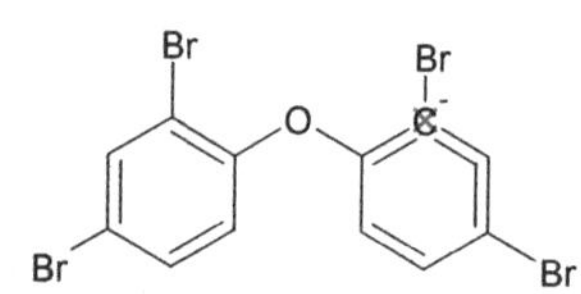

Tetrabromodiphenyl ether

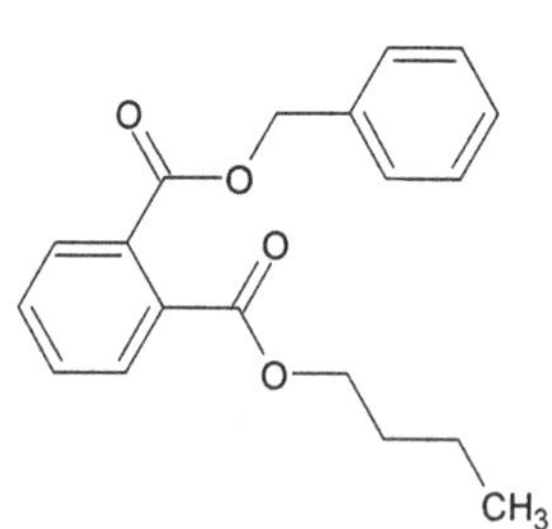

Benzylbutyl phthalate

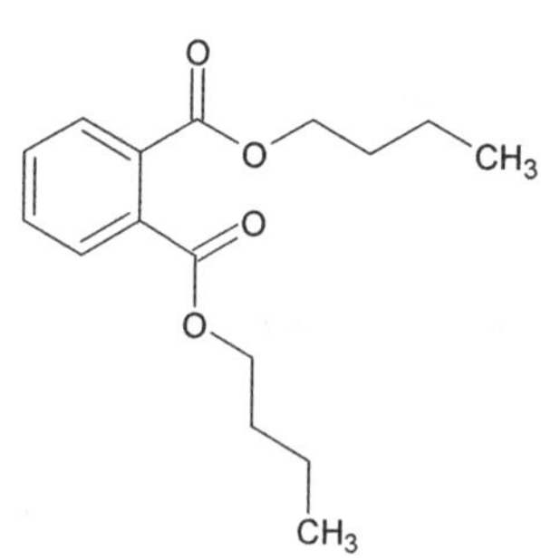

Dibutyl phthalate

Figure 11.1 Hazardous additives (10).

to replace non-sustainable polymers in demanding applications. These chain extenders are also able to compatibilize biopolymers with traditional engineering plastics (13).

11.4.1 *Multiblock Polymer*

Hydrocarbon-based membranes can extend the operating current range of a polymer electrolyte membrane water electrolyzer (PEMWE) because of their low gas permeability and high proton conductivity (14). However, increasing the ion exchange capacity for sufficiently high proton conductivity leads to a poor mechanical stability and an increased gas permeability for hydrocarbon-based membranes.

A poly(*p*-phenylene)-based multiblock polymer was presented with an oligomeric chain extender to get an improved mechanical stability and an extended operating current range. With the introduction of a chain extender, the mechanical and chemical stability were notably enhanced. The effects of ion exchange capacity on mechanical properties, proton conductivity, and hydrogen crossover were investigated compared to a Nafion membrane. The oligomeric chain extender-based PEMWE has a 2.1 times wider operating current range than a Nafion-based PEMWE because of its smaller ohmic overvoltage and lower gas crossover at an optimal ion exchange capacity, demonstrating the possibility of a new hydrocarbon-based membrane for a PEMWE (14).

11.4.2 *Chain Extenders for Polyurea Polymer*

The synthesis of versatile poly(urea) and poly(urethane) matrices, including different chain extenders, was described (15). This process facilitates the design of distinct, tunable properties, and high-performance derivatives.

The polymers can be used for various defense and security applications, such as coatings for ballistic protection, CBRN protection, binders for energetic formulations, and others. Combining aliphatic and aromatic molecules in poly(urea) or poly(urethane) structures enables the synthesis of polymers with improved and controllable thermo-mechanical properties. Thus, for poly(urea) synthesis, several materials can be used as chain extenders. Two types of polymeric aliphatic diamines and three types of aromatic chain extenders were used. These compounds are presented in Table 11.2. Some compounds are shown in Figure 11.2.

Table 11.2 Chain extenders for polyurea (15).

Compound	Compound
1,4-Butanediol	1,3-Butanediol
1,1′-Biphenyl-4,4′-diamine	Benzene-1,2-diamine
1,2-Diphenylhydrazine	Benzene-1,3-diol
Benzene-1,4-diol	Benzene-1,2,3-triol
Ethylene diamine	Isophorone diamine

Recent research has indicated that aromatic chain extenders are significantly more reactive than their aliphatic equivalents, resulting in noticeably shorter gel times (15).

11.5 Nucleating Additives

Compounds and compositions comprising specific metal salts of hexahydrophthalic acid (HHPA) in order to provide highly desirable properties within thermoplastic articles are described in a study (16). These HHPA derivatives are useful as nucleating and/or clarifying agents for such thermoplastics. They are practical and easy to handle.

Such compounds provide excellent crystallization temperatures, stiffness, and an acid scavenger compatibility within poly(olefin)s. Also, such compounds exhibit very low hygroscopicity and therefore excellent shelf stability as powdered or granular formulations. Thermoplastic additive compositions and methods of producing polymers with such compounds are also described (16). Some well-known nucleator compounds are shown in Table 11.3 and in Figure 11.3.

Such compounds impart high polyolefin crystallization temperatures. The crystallization at higher temperatures results in reduced fabrication cycle times and a variety of improvements in physical properties, such as stiffness.

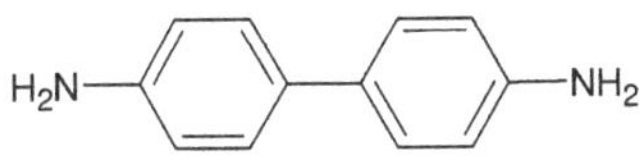

1,1′-Biphenyl-4,4′-diamine

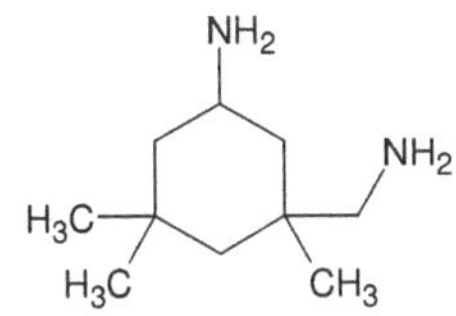

Isophorone diamine

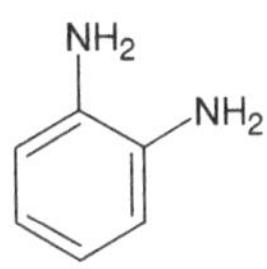

Benzene-1,2-diamine

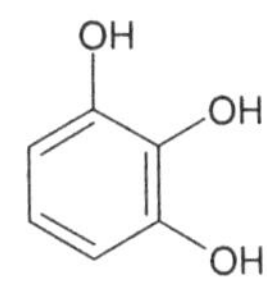

Benzene-1,2,3-triol

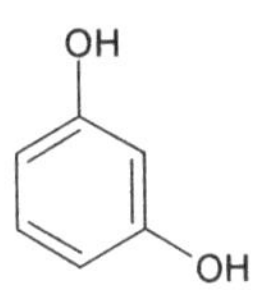

Benzene-1,3-diol

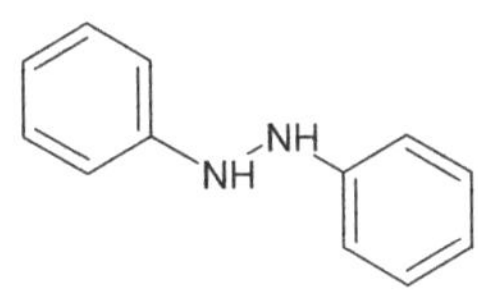

1,2-Diphenylhydrazine

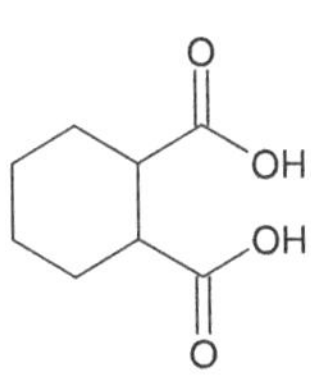

Hexahydrophthalic acid

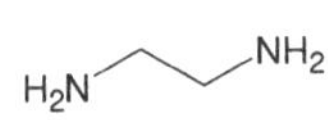

Ethylene diamine

Figure 11.2 Chain Extenders for Polyurea (15).

Table 11.3 Nucleating additives (16).

Compound
Sodium benzoate
1,3-O-2,4-Bis(3,4-dimethylbenzylidene) sorbitol
Sodium 2,2′-methylene-bis-(4,6-di-tert-butylphenyl) phosphate
Aluminum bis[2,2′-methylene-bis-(4,6-di-tert-butylphenyl)phosphate]

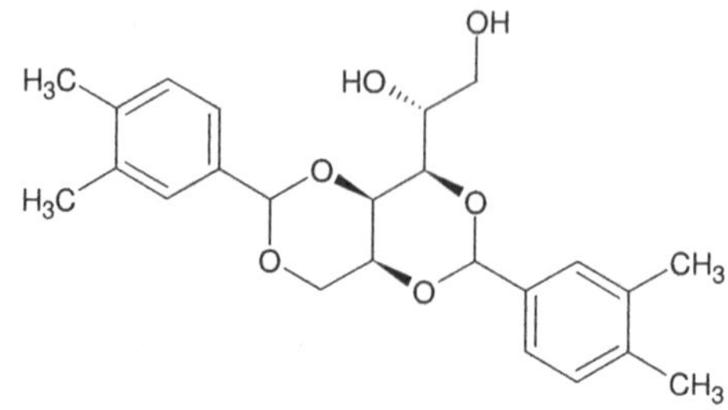

1,3-O-2,4-Bis(3,4-dimethylbenzylidene) sorbitol

Figure 11.3 Nucleating additive for polyurea (16).

11.6 Food Additives

Materials for food additives are listed in a monograph (17). Also, the chemistry of food additives has been described in a monograph (18).

A sensitive, accurate and headspace-solid phase microextraction-gas chromatography-mass spectrometry (HS-SPME-GC-MS/MS) was developed for the determination of styrene in various food matrices (19). The mean recovery ranged from 90% to 116% with a relative standard deviation of =11%.

This method was used for the determination of the concentration of styrene in 23 foodstuffs packed in poly(styrene) (PS) containers, as well as the levels of styrene migrating into various foods (water, milk, cheese or cream) from 14 tableware or kitchenware articles made of styrene plastics.

All the samples were collected from the Greek market in 2020. Styrene concentrations in the packaged foods ranged from 0.4 to 160 $ng\,g^{-1}$ with the highest concentration found in a meat product packed in a foamed PS tray. It is worth noting that 56% of PS packaged dairy products and desserts had a styrene concentration higher than 10 $ng\,g^{-1}$. Particularly high levels of styrene that have not previously been reported, up to 46 $ng\,g^{-1}$, were found in dairy products for children. The highest level of styrene migration from tableware or kitchenware articles, 89 $ng\,g^{-1}$, was observed when disposable cups from foamed PS were filled with milk at 70°C for 2 *h*.

The implications of these findings for the assessment of the potential exposure of the Greek consumers to styrene migrating from PS food contact materials were discussed (19).

11.6.1 *Mycotoxins*

Mycotoxins, chemical substances produced by ubiquitous fungi, can make the difference between profit and loss to the poultry and livestock industries. Animals are extremely vulnerable to mycotoxins due to the common practice of diversion of mycotoxin contaminated agricultural commodities to animal feed (20).

A method has been presented for inactivating mycotoxins which may be present as contaminants in animal feeds by adding a phyllosillicate/sequestrant formulation to animal feed. This will function as a mycotoxin inactivating agent (20).

The most studied and prevalent agents are the aflatoxins, a group of closely related polysubstituted coumarin derivatives, which are biosynthesized by flavis and parasiticus species of *Aspergillus fungi*. The aflatoxins have invoked much concern as toxic food and feedborne agents following the discovery that they (20):

1. are potent carcinogens and mutagens,
2. are stable in foods and feeds and are relatively unaffected by a variety of processing procedures,
3. can be found as residues in the tissues of animals and humans, and
4. are associated with animal and human disease.

A preponderance of poultry and livestock exposure to aflatoxins is chronic in nature and occurs through the ingestion of low levels of these chemicals such as marginally contaminated rations which do not increase the mortality rate nor result in obvious signs of disease. Instead, chronic exposure to aflatoxins results in economically important effects in animals such as depression of growth rates, feed conversion, and alteration of immunocompetency, which can result in increased susceptibility to infection and decreased ability to resist stress.

Numerous approaches to reduce the aflatoxin levels in agricultural commodities have been experimentally assessed. These include mixing and dilution with aflatoxin-free grains in order to obtain a level within regulatory guidelines, i.e., 20 *ppb* or less; physical methods of separation such as cleaning, density segregation and preferential fragmentation; solvent extraction; biological inactivation; thermal inactivation; and chemical inactivation with a variety of acids, aldehydes, oxidizing agents and alkalies (20).

11.6.2 Stabilizers

Flaxseed is one of the main cultivated economic oil crops in the world, with a global output of around 3.06 *Mt* (21). Its residual meal, after oil industrial processing, is rich in protein (35% – 40%). Flaxseed protein is mainly used for feed and its processing efficiency is low. The potential for high-value utilization of flaxseed protein is far from being excavated.

Interface dominated-colloidal systems, such as emulsions and foams, are widely available and have extensive applications in food products. Flaxseed protein possesses high contents of hydrophobic amino acids and sulfur-containing amino acids, relatively ordered secondary structures, tunable structural properties, and acceptable interfacial behaviors. These properties result in the potential to effectively stabilize food emulsions and foams (21).

By modifications of flaxseed proteins, such as physical treatment and the complexation with polysaccharides and polyphenols, their structural properties and interfacial behaviors can be regulated, thereby leading to the enhancement of emulsifying and foaming properties (21).

11.6.3 Microbial Stabilizers

Food stabilizers can help the food retain its structure, improve its water-binding capacity, mouthfeel, thickening, consistency, and viscosity (22).

They are obtained from diverse natural raw resources, viz. land and sea plants, connective tissues of animals, and microorganisms. Among these, microorganisms are receiving momentous attention as a stable and sustainable source of food stabilizers like xanthan gum, gellan, pullulan, and curdlan.

Different kinds of microbes, including bacteria, yeast, and actinomycetes. have the capability to provide access to a larger range of food stabilizers. Also, genetically engineered microbes, viz. *Actinobacillus succinogenes, Basfia succiniciproducens, Mannheimia succiniciproducens, Escherichia coli,* etc., produce diverse stabilizers like amino and carboxylic acids as a part of their carbon's metabolism (22).

11.6.4 Thickeners

Food thickening agents are widely used to modify the rheological and the textural properties as well as to enhance the quality attributes (23).

The improvement in the moisture binding capacity, structural modification and altering flow behavior properties are the major functions of food thickeners.

Modified starches, proteins, individually or in combination with exudates and seed gums, seaweed extracts and, most recently, microbial polysaccharides, are found to have the ability to improve product mouthfeel, handling properties, and stability characteristics. Factors such as temperature, shear, pH, ionic strength, etc., have effects on the functionality of these thickening agents and must be carefully optimized by food processors while formulation (23).

Moreover, the type of thickener utilized has an impact on the product functionality. The addition of a protein-based thickener boosts the amino acid content (arginine, cysteine, histidine, lysine, proline and γ-aminobutyric acid) and favors the nutritional and rheological properties. Some protein-based thickeners are shown in Figure 11.4.

However, realigning the polymer chains reduces the interaction between the adjacent molecules and results in decreased viscosity. Steric stabilization or network formation by the bacterial cellulose fibrils are possible factors for production of stable emulsions (23).

11.6.5 Gelling Agents

Gelling agents for food system applications have been described (24).

The most frequent gelling components used in meals are polysaccharides and proteins. The method of gel formation is determined by the type of the gelling agents and the circumstances of gelation,

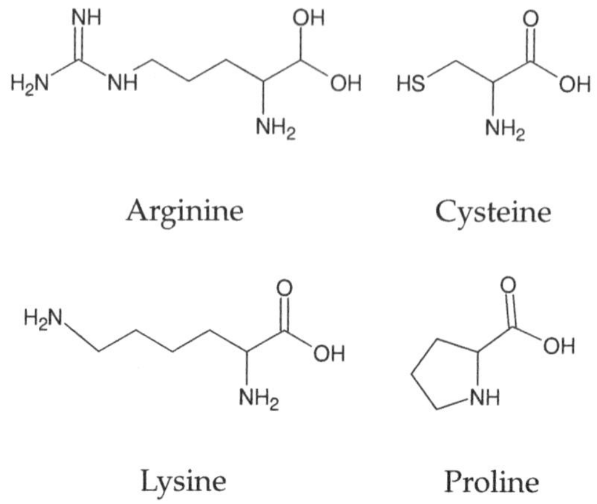

Figure 11.4 Protein-based thickeners (23).

such as temperature, ion presence, pH, and concentration of gelling agents.

Food gels may exceed synthetic alternatives owing to their renewable supply, affordability, bio-compatibility, and bio-degradability. In the food industry, the quality of food products, texture, taste, shelf life, and delivery can be improved through the use of gelling agents, microgels and nanogels. Also, gelling agents can serve as a natural barrier between food and its environment, thus allowing moisture, gases, and volatiles to pass through while conferring protection and controlled release property (24).

Gums and polymers can be found abundantly in nature from various plants, microbial and marine sources. Their physical and functional characteristics can be customized according to their application in product development. The gelling behavior of natural gums and their formulation prospects have been described (25).

This gelling behavior was compared with commercially used polymers, such as Carbopol 934, carboxy methyl cellulose, xanthan, and sodium alginate.

The analysis of these gel formulations indicated that formulations developed using guar gum possess a much better gelling behavior, i.e., viscosity and muco-adhesiveness, then the other selected gums (25).

11.7 Antioxidants

Antioxidants are used to retard the reaction of organic materials with atmospheric oxygen (26). Such a reaction can cause the degradation of the properties of a polymer, loss of flavor and development of rancidity in foods, and a change in lubricants.

The need for antioxidants depends on the chemical composition of the substrate and the conditions of exposure. High concentrations of antioxidants, i.e., 0.5% to 2% based on the weight of he polymer, are used to stabilize polymers such as natural rubber and polyunsaturated oils.

Saturated polymers have a greater oxidative stability and require relatively low concentrations of stabilizers, i.e., 0.01% to 0.05%, based on the weight of the polymer.

The mechanism by which an organic material undergoes autoxidation involves a free-redical chain reaction. High concentrations are also required in substrates that are highly sensitive to oxidation, such as unsaturated elastomers and acrylonitrile-butadiene-styrene.

Thermally induced homolytic decomposition of peroxides and hydroperoxides to free radicals increases the rate of oxidation. The decomposition to nonradical species removes hydroperoxides as potential sources of oxidation initiators.

The selection of an antioxidant or system of antioxidants is dependent on the polymer and the anticipated end use. Low concentrations of antioxidants (< 0.01%) are added to poly(ethylene) and poly(propylene) after synthesis and prior to isolation to retard oxidation of the polymers exposed to air. Polyunsaturated elastomers are sensitive to oxidation. Antioxidants are added to the elastomers prior to vulcanization to protect the rubber during drying and storage. Nonstaining antioxidants may be used.

Staining antioxidants are preferred for the manufacture of tires. Gasoline and jet engine fuels contain unsaturated compounds that oxidize during storage, darken, and form gums and deposits. Radical scanvengers and alkylated paraphenylene diamines are used as stabilizers.

Lubricants for gasoline engines are required to withstand harsh conditions. Relatively high concentrations of primary antioxidants and synergists are used to stabilize lubricating oils. The oxidation of food containing fats and oils results in a loss of sensory appeal and eventually rancidness. The U.S. Food and Drug Administration (FDA) regulations covering the use of direct food additives are stringent and few new materials have been regulated.

The most commonly used materials are butylated hydroxytoluene *n*-propyl gallatte, α-tocopherol, and butylated hydroxy anisole. A number of antioxidants have been accepted by the FDA as indirect additives for polymers used in food applications (26).

The use of synthetic and natural antioxidants in the stabilization of food products have been reviewed in a chapter of a book (27). Here, both natural and synthetic antioxidants were categorized in terms of how they interfere with the oxidation process. Also, new areas for the development of more effective natural food antioxidants were suggested by examining and contrasting the latest antioxidant technology now in use in the related field of polymer science.

11.7.1 *Natural Antioxidants*

Natural antioxidants as stabilizers for polymers have been detailed (28). Natural antioxidants are used in plastic products during their processing and application. A classification based on the mechanism of action is shown in Table 11.4.

Table 11.4 Classification of non-enzymatic natural antioxidants by their mechanism (28).

Type	Mechanism	Reacts with	Examples
Carotenoids	quenching, electron transfer, radical adduct formation	singlet oxygen, alkyl, alkoxy, peroxy radicals	carotenes, xanthophylls
Polyphenols	H-atom transfer, transition metal chelation	alkoxy, peroxy radicals, transition metals	flavonoids, curcuminoids, tocopherols, phenolic acids
Phenolic polymers	H-atom transfer, transition metal chelation	alkoxy, peroxy radicals, transition metals	lignin, tannin

The antioxidants presented in Table 11.4 can be found in seeds, leaves of plants, berries, peels of fruits and vegetables, since these species are constantly exposed to ultraviolet radiation, which demand a high level of antioxidants (29). The main sources of carotenoids are sweet potato (30), carrots (31), tomato (32), peppermint and spearmint (33). In addition, spinach, corn (34) and kale (35) are rich in xanthophylls.

Flavonoids are the largest group of natural phenolic antioxidants with more than 6000 representatives (36). They can be found in the skin of red grapes (37, 38) and in red wine (38), in various citruses (39), onion (40) and honey (41), but leaves of stevia (42) and green tee contain them in considerable amounts too (43).

Various seeds and spices like mustard, ginger, fennel and pepper contain large amounts of phenolic acids (44), and curcuminoids can be found in turmeric (31). Lignin is the major component of the cell wall of vascular plants (45). Tannins can be found in the seed and skin of grapes (46) and in other exotic fruits (47), but also in the leaves and barks of woods (48).

11.7.2 Polyphenolic Antioxidants

Antioxidants are widely known as natural components of healthy food and drinks or as additives to commercial polymer materials to prevent their degradation (49).

An increasing interest has occurred for enhancing the antioxidant functionality of newly developed polymer materials and coatings.

A comparative analysis of multiple ways of integrating antioxidants within diverse polymer materials, including bulk films, electrospun fibers, and self-assembled coatings, has been shown.

Polyphenolic antioxidant moieties with varied molecular architecture are the focus of study, because of their abundance, nontoxic nature, and potent antioxidant activity (49). Polymer materials with integrated polyphenolic functionality offer opportunities and challenges that span from the fundamentals to their applications.

In addition to the traditional blending of antioxidants with polymer materials, developments were made in surface grafting and assembly via noncovalent interaction for controlling localization versus migration of antioxidant molecules. The versatile chemistry of polyphenolic antioxidants offers numerous possibilities for programmed inclusion of these molecules in polymer materials using not only van der Waals interactions or covalent tethering to polymers, but also via their hydrogen-bonding assembly with neutral molecules.

An understanding and rational use of interactions of polyphenol moieties with surrounding molecules can enable a precise control of concentration and retention versus delivery rate of antioxidants in polymer materials that are critical in food packaging, biomedical, and environmental applications (49).

11.7.3 Mango Peel

The application of mango peel as an antioxidant additive for an edible oil has been described (50). Mango peel, an important by-product of industrial mango processing, is rich in phytochemical compounds. The operating parameters of the supercritical CO_2 extraction of carotenoids from mango peel by means of a Box-Behnken design were optimized.

The mango peel was characterized in terms of its proximate composition. Furthermore, the effects of freeze-drying and air-drying on its total carotenoid content were investigated (50). The carotenoids profile of the optimal supercritical CO_2 extract was characterized by RP-UHPLC-DAD. This extract was evaluated as natural antioxidant for sunflower oil.

Mango peel displayed high contents of crude protein and phenolic compounds. Although freeze-drying prevented the degradation of all-trans-β-carotene, air-drying was selected for the valorization strategy. The higher yields of carotenoids were found at 25.0*MPa*, 60°C, and 15% ethanol. Under these conditions the mango peel extracts protected the sunflower oil against lipid oxidation (50).

11.8 Future Trends in Summary

This chapter explained several additive classes for polymers.

Examples of additive types are compatibilizers, contaminants, legacy additives, chain extenders, nucleating additives, and food additives.

References

1. K. Ragaert, L. Delva, and K. Van Geem, Mechanical and chemical recycling of solid plastic waste, *Waste Management* 24–58 (2017).
2. G.H. Fredrickson, S. Xie, J. Edmund, M.L. Le, D. Sun, D.J. Grzetic, D.L. Vigil, K.T. Delaney, M.L. Chabinyc, and R.A. Segalman, Ionic compatibilization of polymers, *ACS Polymers Au* 299–312 (2022).
3. L.A. Utracki, D.J. Walsh, and R.A. Weiss, Polymer alloys, blends, and ionomers: An overview (1989).
4. G.P. Simon, *Polymer Blends and Alloys*, Routledge, (2019).
5. W.E. Baker, C.E. Scott, G.-H. Hu, and M. Akkapeddi, *Reactive Polymer Blending*, Vol. 1, Hanser Munich, (2001).
6. M.C. Liu and H. Huang, *Types of Reactive Polymers Used in Blending*, Vol. 1 chapter 2, 13–42. Hanser Munich, (2001).
7. A. Eisenberg, P. Smith, and Z.-L. Zhou, Compatibilization of the polystyrene/poly (ethyl acrylate) and polystyrene/polysoprene systems through ionic interactions, *Polymer Engineering & Science* 1117–1122 (1982).
8. P. Agarwal, I. Duvdevani, D. Peiffer, and R. Lundberg, Polymer blend complexes based on coordination chemistry, *Journal of Polymer Science Part B: Polymer Physics* 839–854 (1987).
9. T.P. Russell, R. Jerome, P. Charlier, and M. Foucart, The microstructure of block copolymers formed via ionic interactions, *Macromolecules* 1709–1717 (1988).
10. S. Wagner and M. Schlummer, Legacy additives in a circular economy of plastics: Current dilemma, policy analysis, and emerging countermeasures, *Resources, Conservation and Recycling* 104800 (2020).
11. P.K. Rai, C. Sonne, R.J. Brown, S.A. Younis, and K.-H. Kim, Adsorption of environmental contaminants on micro-and nano-scale plastic polymers and the influence of weathering processes on their adsorptive attributes, *Journal of Hazardous Materials* 127903 (2022).
12. N. Morin-Crini, E. Lichtfouse, M. Fourmentin, A.R.L. Ribeiro, C. Noutsopoulos, F. Mapelli, É. Fenyvesi, M.G.A. Vieira, L.A. Picos-Corrales, J.C. Moreno-Piraján, et al., Removal of emerging contaminants from wastewater using advanced treatments. A review, *Environmental Chemistry Letters* 1333–1375 (2022).
13. V. Frenz, D. Scherzer, M. Villalobos, A.A. Awojulu, M. Edison, and R. Van Der Meer, Multifunctional polymers as chain extenders and compatibilizers for polycondensates and biopolymers, in *Tech. Pap. Reg. Tech. Conf. Soc. Plast. Eng*, Vol. 3 1678–1682. Citeseer, (2008).
14. S. Choi, S.-H. Shin, D.-H. Lee, G. Doo, D.W. Lee, J. Hyun, S.H. Yang, D.M. Yu, J.Y. Lee, and H.-T. Kim, Oligomeric chain extender-derived poly (p-phenylene)-based multi-block polymer membranes for a wide

operating current density range in polymer electrolyte membrane water electrolysis, *Journal of Power Sources* 231146 (2022).
15. G. Toader, A.E. Moldovan, A. Diacon, F.M. Dirloman, E. Rusen, A. Podaru, T. Rotariu, R.E. Ginghina, and O.E. Hoza, Effect of aromatic chain extenders on polyurea and polyurethane coatings designed for defense applications, *Polymers* 756 (2023).
16. D.L. Dotson and X.E. Zhao, Metals salts of hexahydrophthalic acid as nucleating additives for crystalline thermoplastics, US Patent 6 599 971, assigned to Milliken and Co., July 29, 2003.
17. A.L. Branen, P.M. Davidson, S. Salminen, and J. Thorngate, (Eds.), *Food Additives*, Food Science and Technology, CRC Press, Boca Raton, Florida, 2nd edition, (2001).
18. T.A.M. Msagati, *The Chemistry of Food Additives and Preservatives*, Wiley, (2012).
19. S. Kontou, E. Dessipri, and E. Lampi, Determination of styrene monomer migrating in foodstuffs from polystyrene food contact articles using HS-SPME-GC-MS/MS: Results from the Greek market, *Food Additives & Contaminants: Part A* 415–427 (2022).
20. D.R. Taylor, K.A. Delaney, and T.D. Phillips, Animal feed additive and method for inactivating mycotoxins present in animal feeds, US Patent 5 165 946, assigned to Texas A&M University System, Engelhard Corp., November 24, 1992.
21. D. Peng, J. Ye, W. Jin, J. Yang, F. Geng, and Q. Deng, A review on the utilization of flaxseed protein as interfacial stabilizers for food applications, *Journal of the American Oil Chemists' Society* 723–737 (2022).
22. B. Devi, S. Devi, D. Sood, and P. Arya, Microbial stabilizers in food processing in *Microbes for Natural Food Additives* 113–145. Springer, (2023).
23. P. Himashree, A.S. Sengar, and C. Sunil, Food thickening agents: Sources, chemistry, properties and applications-A review, *International Journal of Gastronomy and Food Science* 100468 (2022).
24. N. Munir, M. Hasnain, H. Waqif, B.O. Adetuyi, C. Egbuna, M.C. Olisah, C.J. Chikwendu, C.Z. Uche, K. C. Patrick-Iwuanyanwu, and A.M.A. El Sayed, Gelling agents, micro and nanogels in food system applications in *Application of Nanotechnology in Food Science, Processing and Packaging* 153–167. Springer, (2022).
25. R. Mohan, S. Singh, G. Kumar, and M. Srivastava, Evaluation of gelling behavior of natural gums and their formulation prospects, *Indian J. Pharm. Educ. Res* 1016–1023 (2020).
26. R. Thomas, M. Dexter, and R. King III, Antioxidants, polymers, *Kirk-Othmer Encyclopedia of Chemical Technology* (2000).
27. D. Berdahl, R. Nahas, and J. Barren, Synthetic and natural antioxidant additives in food stabilization: Current applications and future research in E.A. Decker, (Ed.), *Oxidation in Foods and Beverages and*

Antioxidant Applications, Woodhead Publishing Series in Food Science, Technology and Nutrition chapter 12, 272–320. Woodhead Publishing, (2010).

28. B. Kirschweng, D. Tátraaljai, E. Földes, and B. Pukánszky, Natural antioxidants as stabilizers for polymers, *Polymer Degradation and Stability* 25–40 (2017).
29. P. Majer, S. Neugart, A. Krumbein, M. Schreiner, and É. Hideg, Singlet oxygen scavenging by leaf flavonoids contributes to sunlight acclimation in Tilia platyphyllos, *Environmental and Experimental Botany* 1–9 (2014).
30. S. Laurie, P. Van Jaarsveld, M. Faber, M. Philpott, and M. Labuschagne, Trans-β-carotene, selected mineral content and potential nutritional contribution of 12 sweetpotato varieties, *Journal of Food Composition and Analysis* 151–159 (2012).
31. S. Gautam, K. Platel, and K. Srinivasan, Influence of β-carotene-rich vegetables on the bioaccessibility of zinc and iron from food grains, *Food Chemistry* 668–672 (2010).
32. R. Ilahy, C. Hdider, M.S. Lenucci, I. Tlili, and G. Dalessandro, Antioxidant activity and bioactive compound changes during fruit ripening of high-lycopene tomato cultivars, *Journal of Food Composition and Analysis* 588–595 (2011).
33. A. Curutchet, E. Dellacassa, J.A. Ringuelet, A.R. Chaves, and S.Z. Viña, Nutritional and sensory quality during refrigerated storage of fresh-cut mints (Mentha × piperita and M. spicata), *Food Chemistry* 231–238 (2014).
34. M.S. Pasaporte, F.J.R. Rabaya, M.M. Toleco, and D.M. Flores, Xanthophyll content of selected vegetables commonly consumed in the philippines and the effect of boiling, *Food Chemistry* 35–40 (2014).
35. C. Arnold, S. Jentsch, J. Dawczynski, and V. Böhm, Age-related macular degeneration: Effects of a short-term intervention with an oleaginous kale extract. a pilot study, *Nutrition* 1412–1417 (2013).
36. J.B. Harborne and C.A. Williams, Advances in flavonoid research since 1992, *Phytochemistry* 481–504 (2000).
37. C. Alcalde Eon, I. García Estévez, A. Martín Baz, J.C. Rivas Gonzalo, M.T. Escribano Bailón, et al., Anthocyanin and flavonol profiles of Vitis vinifera L. cv Rufete grapes, *Biochemical Systematics and Ecology* (2014).
38. N.-N. Liang, B.-Q. Zhu, S. Han, J.-H. Wang, Q.-H. Pan, M.J. Reeves, C.-Q. Duan, and F. He, Regional characteristics of anthocyanin and flavonol compounds from grapes of four Vitis vinifera varieties in five wine regions of China, *Food Research International* 264–274 (2014).
39. H.R. de Moraes Barros, T.A.P. de Castro Ferreira, and M.I. Genovese, Antioxidant capacity and mineral content of pulp and peel from commercial cultivars of citrus from Brazil, *Food Chemistry* 1892–1898 (2012).

40. S. Kühn, H. Wollseifen, R. Galensa, N. Schulze-Kaysers, and B. Kunz, Adsorption of flavonols from onion (Allium cepa L.) processing residues on a macroporous acrylic resin, *Food Research International* 103–108 (2014).
41. J.M. Alvarez-Suarez, F. Giampieri, A.M. González-Paramás, E. Damiani, P. Astolfi, G. Martinez-Sanchez, S. Bompadre, J.L. Quiles, C. Santos-Buelga, and M. Battino, Phenolics from monofloral honeys protect human erythrocyte membranes against oxidative damage, *Food and Chemical Toxicology* 1508–1516 (2012).
42. A. Periche, M.L. Castelló, A. Heredia, and I. Escriche, Influence of drying method on steviol glycosides and antioxidants in Stevia rebaudiana leaves, *Food Chemistry* 1–6 (2015).
43. I.K. Bae, H.M. Ham, M.H. Jeong, D.H. Kim, and H.J. Kim, Simultaneous determination of 15 phenolic compounds and caffeine in teas and mate using RP-HPLC/UV detection: Method development and optimization of extraction process, *Food Chemistry* 469–475 (2015).
44. M. Lu, B. Yuan, M. Zeng, and J. Chen, Antioxidant capacity and major phenolic compounds of spices commonly consumed in China, *Food Research International* 530–536 (2011).
45. Z. Ji, J.-F. Ma, Z.-H. Zhang, F. Xu, and R.-C. Sun, Distribution of lignin and cellulose in compression wood tracheids of Pinus yunnanensis determined by fluorescence microscopy and confocal Raman microscopy, *Industrial Crops and Products* 212–217 (2013).
46. J.F. Harbertson, G.P. Parpinello, H. Heymann, and M.O. Downey, Impact of exogenous tannin additions on wine chemistry and wine sensory character, *Food Chemistry* 999–1008 (2012).
47. F. Luo, Y. Fu, Y. Xiang, S. Yan, G. Hu, X. Huang, G. Huang, C. Sun, X. Li, and K. Chen, Identification and quantification of gallotannins in mango (Mangifera indica L.) kernel and peel and their antiproliferative activities, *Journal of Functional Foods* 282–291 (2014).
48. E. Onem, G. Gulumser, M. Renner, and O. Yesil-Celiktas, High pressure vegetable tanning of sheepskins using supercritical carbon dioxide, *The Journal of Supercritical Fluids* 259–264 (2015).
49. J. Brito, H. Hlushko, A. Abbott, A. Aliakseyeu, R. Hlushko, and S.A. Sukhishvili, Integrating antioxidant functionality into polymer materials: fundamentals, strategies, and applications, *ACS Applied Materials & Interfaces* 41372–41395 (2021).
50. A. del Pilar Sanchez-Camargo, L.-F. Gutierrez, S.M. Vargas, H.A. Martinez-Correa, F. Parada-Alfonso, and C.-E. Narvez-Cuenca, Valorisation of mango peel: Proximate composition, supercritical fluid extraction of carotenoids, and application as an antioxidant additive for an edible oil, *The Journal of Supercritical Fluids* 104574 (2019).

12

Manufacturing

12.1 Wood-Plastic Composites

Wood-plastic composites (WPCs) have shown promising domestic and industrial applications. Both virgin and waste materials have been used as matrix and reinforcement materials in WPCs manufacturing, which enhances the eco-efficiency of WPCs products (1).

12.1.1 Fabrication

The WPCs manufacturing techniques, including pre-processing of the composite ingredients and post-processing of the WPCs products, were reported.

Here, three main types of pre-processing techniques were discussed; chemical, mechanical, and thermal treatments. Conventional manufacturing techniques of WPCs, such as extrusion molding, injection molding, compression molding, stir casting, and hot-press method, have also been presented. In addition, advanced manufacturing techniques, such as selective laser sintering ,have been discussed (1). Also, the recycling of WPC is discussed (1).

A WPC is a composite material made of plastic as a matrix and wood as filler (2)

Like other composite materials, the constituent materials are preserved in their original forms and are incorporated to obtain a new composite material with reasonable mechanical and physical properties and low cost. It is formed in plank or beam shape which may be utilized in many applications such as outdoor deck floors, railings, park benches, car door panels, car seat backs, fences, door and window frames, timber plate structures, and indoor furniture (3). Furthermore, they have shown promising applications as thermal and sound insulation panels (4).

12.1.2 Manufacturing Processes

Current manufacturing techniques for thermoplastic-fiber composite materials focus on the extrusion process, whereby a molten cellulose/thermoplastic mix is forced under high pressure through a forming die (5). The product is shaped and formed corresponding to the die, and then allowed to cool. In some instances, the cooling of the extruded product is accomplished by wrapping the extruded product around one or more large, chilled, cylindrical mandrels. Such a cooling technique, however, flexes the product, resulting in unacceptable fracturing of sheet or panel product having a thickness of more than about 0.5 inches.

Alternatively or additionally, known fabrication processes require post-extrusion cooling mechanisms that are dozens of feet long, demanding large manufacturing facilities. Known extrusion-type production devices generally also are limited as to both the width and thickness of panels or sheets than can effectively be produced. A further shortcoming of known systems and methods for manufacturing cellulose/plastic composites is the difficulty in fabricating a composite panel to which a polymeric film may be easily and securely attached with an ordinary known, and perhaps yet-to-be-developed, adhesive.

Known extrusion processes for wood-plastic composite panel fabrication result in a cooled panel with a very smooth surface having thereon a thin layer of melted polymers. This layer or coating is an unavoidable consequence of the heated extrusion process for extruding a wood-plastic composite. The thin layer manifests a slick, oily, character that resists adhesion thereto of most commonly used adhesives, especially adhesives by which, for example, a thin plastic layer or film can be attached to the panel surface.

It is desirable to secure to a composite panel surface another layer, such as a thin plastic film, when, for example, it is desired to apply a visual sign or other layer to the cooled panel surface. Extruded wood-plastic composites exhibit excellent durability, strength, and weathering characteristics for use as the foundation for road signs, but many preferred types of film applications used as signs are difficult to apply to such composites (5).

A production process is shown in Figure 12.1.

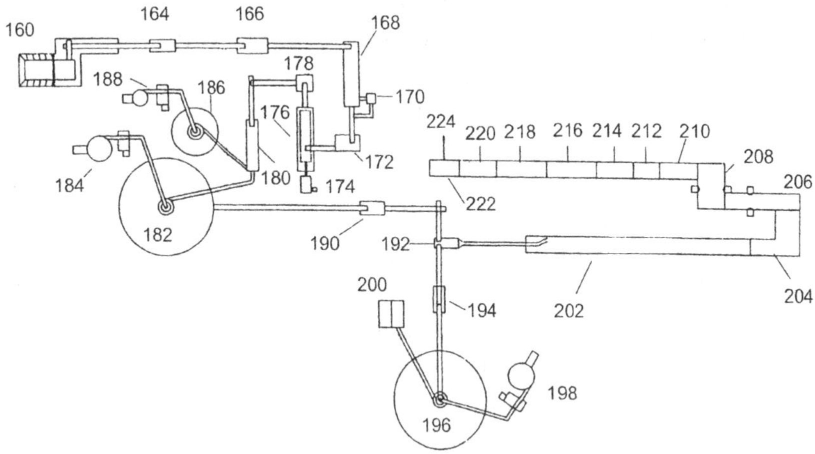

Figure 12.1 Production process (6).

As shown in Figure 12.1, a wood receiver 160 is used for receiving wood. Next, the wood is transported to a screen 164 for screening out undesirable larger pieces of wood. Once screened, the wood is transported to a reducer 166 for reducing the size of the screened wood. The reduced wood is transported to a washer and/or screener 168 that optionally includes a re-chipper 170 for further reduction of screened wood.

Next, the washed wood is flaked using a flaker 172. From the flaker 172, the wood is transported to a flake receiver 176 that optionally contains a heater and/or dryer and/or a dust burner 174. Flakes are then transported to a grinder 178 for grinding and/or sizing of the wood. The ground wood is then transported to a sifter 180 for sifting fines from larger pieces of wood. Fines are transported to and stored in a fines receiver 186 while the larger pieces are transported to and stored in a processed wood receiver 182. The fines receiver and processed wood receiver optionally includes filters 184, 188. The processed wood and/or fines are then ready for further processing and/or combination with plastic.

Plastic enters the production process through a plastic receiver 200. The plastic in the plastic loader is optionally transported to a plastic storage receiver 196 for storage and/or further processing. The plastic storage receiver 196 optionally includes a filter 198.

Once the wood has been processed and/or stored, it is then transported to a measurement system 190, for example, a weigh station system and/or flow measurement system. Once the plastic has been processed and/or stored, it is then transported to a measurement system 194, for example, a weigh station system and/or flow measurement system. The plastic and wood are transported to a blender 192 for blending plastic and wood.

Blended plastic and wood are transported to a production line 202 that has a press and/or an extruder. The production line produces a final product or optionally includes an additional equipment for performing additional steps for producing a final product. For example, the production line optionally includes an unloader and/or cooler 204; a trimmer and/or borer 206; an additional trimmer and/or borer 208; a transfer and/or inspection unit 210; a sander 212; an additional transfer unit and/or inspection unit 214; a paint unit 216, for example, for spray painting; an oven 218, for example, for curing paint and/or other material; a grade station 220; and/or a stacker 222, for stacking product. Product output from the production line is output generally at a product or stacked product output 224 (6).

A pressing apparatus for fabricating a dense, rigid, durable panel from a molten composite material has been developed (5). The apparatus features a first platen in spaced and angled relation to a second platen to define a gap between these platens. The gap decreases uniformly from first ends of the platens toward second ends of the platens, and conveyer belts, movably disposed along opposing surfaces of said platens, move molten composite material through the decreasing gap to compress the molten composite material between the platens.

Cooled drums are provided whereby the molten composite is cooled shortly after extrusion. A fine screen, preferably a metal screen mesh, is rolled into one side (surface) of the extruded panel prior to the panel's completed cooling.

The screen is not fully embedded into the extruded panel, rather it remains partially exposed on one side of the panel, and provides that surface with a character amenable to the application of adhesives.

Also, a method has been shown for making a composite wood-plastic composition which incorporates a mesh on a working surface thereof (5). Secondary films and sheets may be readily and securely adhered to the working surface without the need for flaming or similar pre-treating the working surface to remove anti-adherent polymer residues from prior extrusion.

Here, a primary objective is to provide a method and apparatus for manufacturing a cellulose/thermoplastic composite panel with a substantial thickness, width, density and strength, but to a working surface on which a thin film can later be securely adhered using conventional adhesives, and without the need to pre-treat the working surface (5).

12.1.3 Properties

In general, WPCs have many advantages compared with reinforced plastic with mineral fillers such as low cost, low processing problems, in addition to reasonable mechanical properties (7).

Compared to natural wood, WPC exhibits a better dimensional stability in harsh environments (8).

Moreover, WPC has good resistance to termites and fungi (9). The most important issue related to WPC fabrication is its capability to exploit wood and plastic wastes to produce a new product with enhanced properties (10). The main drawback of use of organic fibers in WPC is their tendency to aggregate during processing. In addition, their low moisture resistance results in formation of voids, which impairs the dimensional stability and mechanical properties of the produced WPC. Moreover, the degradation of organic fibers in harsh sunny environments affects the reliability of WPC.

The existing technologies for producing WPC are highly automated and adaptable to various types of raw materials (2). The main conventional processes used in WPC manufacturing are injection molding (11), extrusion (12), and compression molding (13). In addition, new advanced additive manufacturing processes using laser sintering (14) and fused layer modeling (15) have also been proposed. The raw materials of WPCs should undergo pre-processing treatments to enhance the interfacial bonding between the WF and the polymer matrix (16).

12.1.4 Polymeric Materials

In general, most low-cost plastics, such as poly(propylene) (PP), high-density poly(ethylene) (HDPE), low-density poly(ethylene) (LDPE), poly(tetrafluoroethylene), high impact poly(styrene), and poly(styrene) (PS), are characterized as low strength materials (17, 18).

Therefore, mineral fillers, such as calcium carbonate, are added to these polymers during the fabrication process to enhance their mechanical properties (19). However, the rise in the price of calcium carbonate as well as its negative environmental impacts can motivate the discovery of eco-friendly alternatives (20).

Therefore, organic fibers extracted from different types of wood as well as wheat, rice, or flax straw have been proposed as a good alternative to inorganic fillers (18, 21). The extraction of fibers should be done using low-cost environmentally friendly technologies. Wood is considered as the master of all organic fibers due to its strong structure and high mechanical properties compared to other organic fibers (22).

Moreover, the seasonality of annual crops, such as rice, wheat, and flax, is another drawback of these types of organic fibers. The seasonality of these crops increases the cost of the preparation of their fibers due to the special consideration that should be applied during harvesting, drying, separating, transporting, and storing process. Another important issue in fabrication of fiber-polymer composites is the local availability of the organic fibers. Jute fibers may be the optimal choice for fiber-polymer composites in Bangladesh while rice straw fibers are optimal in Egypt (23).

Various polymeric materials (thermosets, thermoplastics) have been used in WPC fabrication (2). The processing temperature of these polymers should not exceed 200°C to protect wood fibers from disintegration of cellulose chains. The polymers that fulfilled these requirements are HDPE, LDPE, poly(vinyl chloride) (PVC), PP, PVC, acrylonitrile-butadiene-styrene, and PS (18). Since HDPE has a good resistance to chemicals, it is used for packing detergents, acids, margarine, and milk. Due to its low melting point, LDPE can be used in heat sealing applications. It is also used in cable and wire applications due to its good electrical characteristics. However, HDPE has a high strength-to-density ratio and high specific strength, which makes it harder than LDPE. HDPE has a good resistance to different solvents. Although PP has similar properties as PE, it has more heat resistance and slightly better mechanical properties. Since PVC has excellent chemical resistance and stability, and good electrical characteristics, it is widely used in fittings and the pipe industry.

However, PVC produces toxic emissions under high temperatures. Due to its high melting point, PP is used in packing applications that are subjected to high temperatures. It is also used in yogurt packing, medicine and ketchup bottles, and the outer casing of car batteries. Acrylonitrile-butadiene-styrene (ABS) has excellent abrasion and chemical resistance, excellent strength and stiffness, is easily formed, machined, and painted, and had reasonable dimensional stability. The good mechanical and physical properties of PS, such as low density, high stiffness and superior thermal insulation, make it a good choice in food and medicine packing applications. Other plastics, such as PBS, PC, PES, PET, PF, PHBV, PLA, PPC, and UPE, have been used as matrix in WPCs and showed enhanced mechanical and thermophysical properties.

12.1.5 Recycling

Recycling of plastics can be executed using one of three main techniques: mechanical recycling, source reduction, or feedstock recycling. However, it was reported that the processing of waste material is a cumbersome problem which needs a lot of resources, which impairs its profitability (24). Moreover, most manufacturers are not interested in using waste materials as raw materials in their industries due to the problems related to the material collecting and handling as well as the unstable quality of the waste material (22).

12.2 Single-Use Plastics

12.2.1 *Green Composites*

Palm waste can be used for the production and preparation of green biodegradable composites intended for single-use plastics, such as cutlery and packaging applications, with competitive properties (25). Poly(lactic acid) (PLA) is a suitable polymer that would be compatible with conventional biodegradable polymers to be used in the preparation of date palm biomass-polymer composite.

The preparation of poly(lactic acid)-based composites can be done as follows (25):

Preparation 12–1: A sieved Date Palm Rachis sample (90 μm) was added as a filler to the biodegradable polymer of PLA in three different compositions of 30%, 40%, and 50% by weight. The Date Palm Rachis was not subjected to any chemical treatments or surface modifications to avoid the use of chemicals in food-grade applications, reduce the production cost and save the energy used in cellulose extraction, whereas 30% of extracted cellulose was used to reinforce PLA. Melt mixing was performed using a counter-rotating two-screw extruder at a screw speed of 140 *rpm*. The control sample was prepared by blending pure PLA under the same conditions. Different extruder mixing trials for 30% Date Palm Rachis-PLA composite shows that the best composite color with a recovery of 79.9% and output thickness of 0.15 *cm* was produced by the melting of PLA at 180°C for 7 *min* along with plasticizers up to 10% followed by the addition of Date Palm Rachis to be mixed for 3 *min*. However, the extruded biocomposite was re-melted via injection molding machine in which the homogeneous molten plastic was injected at high-pressure of 500 *bar* to fill the 80°C closed mold cavity of a particular shape in 15 *s* to produce lab test specimens such as dumbbell-shaped specimens.

The possibility of utilizing whole date palm rachis waste without extraction as a filler to produce a composite has been investigated (25). Here, the effect of the addition of biomass (30%, 4%, and 50%) to the PLA matrix on the properties of the resulting composite material, including phase adhesion, degradation temperature, glass transition temperature, and melt flow index, was studied.

The effect with a maximum of 10% of two food-grade plasticizers, i.e., triethyl citrate and poly(butylene adipate terephthalate), in such composites to enhance the mechanical properties of the composite material was studied (25).

The addition of 10% of triethyl citrate showed a superior effect compared with poly(butylene adipate terephthalate) by the improvement in the elongation at break of the optimal composite from 2.10% to 4.20%. Poly(butylene adipate terephthalate) is shown in Figure 12.2 and triethyl citrate is shown in Figure 12.3.

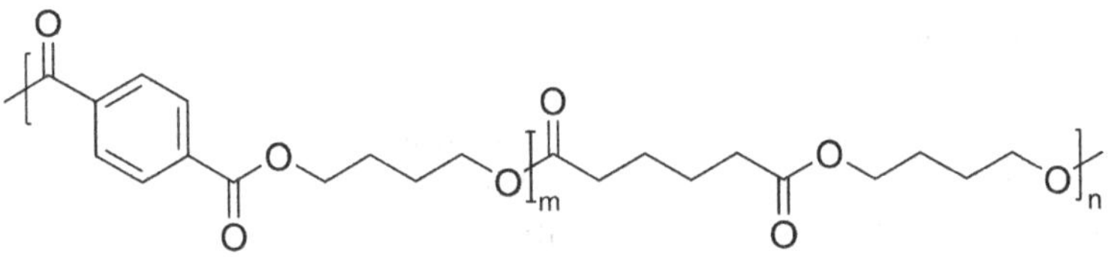

Figure 12.2 Poly(butylene adipate terephthalate).

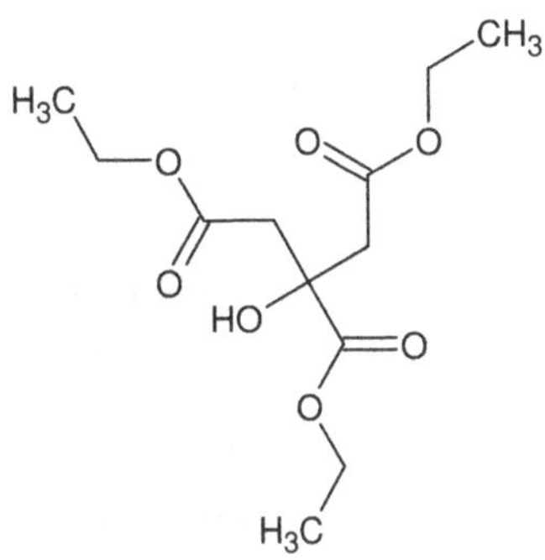

Figure 12.3 Triethyl citrate.

12.2.2 *Diagnostic Waste*

The problem of diagnostic waste goes well beyond the issue of how to replace or dispose of single-use plastics (26). Diagnostic tests often contain chemical reagents that need to be disposed of safely after use, dictated by local chemical waste disposal routes. Illegal or negligent disposal of chemical liquid waste in landfills and public sewage systems can lead to the contamination of groundwater and drinking water.

12.3 Future Trends in Summary

This chapter dealt with some of the issues of some special manufacturing techniques concerning food plastic products and single-use plastics.

References

1. A.H. Elsheikh, H. Panchal, S. Shanmugan, T. Muthuramalingam, A.M. El-Kassas, and B. Ramesh, Recent progresses in wood-plastic composites: Pre-processing treatments, manufacturing techniques, recyclability and eco-friendly assessment, *Cleaner Engineering and Technology* 100450 (2022).
2. D.J. Gardner, Y. Han, and L. Wang, Wood–plastic composite technology, *Current Forestry Reports* 139–150 (2015).
3. M.M.-U. Haque, K. Goda, S. Ogoe, and Y. Sunaga, Fatigue analysis and fatigue reliability of polypropylene/wood flour composites, *Advanced Industrial and Engineering Polymer Research* 136–142 (2019).
4. X. Yang, X. Tang, L. Ma, and Y. Sun, Sound insulation performance of structural wood wall integrated with wood plastic composite, *Journal of Bioresources and Bioproducts* 111–118 (2019).
5. P.T. Archuletta, Method for making a durable wood-plastic composite panel upon which plastic films may be readily and securely adhered, US Patent 9 573 310, assigned to Individual, February 21, 2017.
6. P.T. Archuletta and J.H. Muehl, Method for making wood and plastic composite material, US Patent 6 632 387, assigned to P&M Plastics, US Department of Agriculture USDA, October 14, 2003.
7. S. Rahman, M.N. Islam, S.B. Ratul, N.H. Dana, S.M. Musa, M.O. Hannan, et al., Properties of flat-pressed wood plastic composites as a function of particle size and mixing ratio, *Journal of Wood Science* 279–286 (2018).
8. K.B. Adhikary, S. Pang, and M.P. Staiger, Dimensional stability and mechanical behaviour of wood–plastic composites based on recycled and virgin high-density polyethylene (HDPE), *Composites Part B: Engineering* 807–815 (2008).
9. M. Mankowski and J.J. Morrell, Patterns of fungal attack in wood-plastic composites following exposure in a soil block test, *Wood and Fiber Science* (2000).
10. D. Basalp, F. Tihminlioglu, S.C. Sofuoglu, F. Inal, and A. Sofuoglu, Utilization of municipal plastic and wood waste in industrial manufacturing of wood plastic composites, *Waste and Biomass Valorization* 5419–5430 (2020).

11. P.-Y. Kuo, S.-Y. Wang, J.-H. Chen, H.-C. Hsueh, and M.-J. Tsai, Effects of material compositions on the mechanical properties of wood–plastic composites manufactured by injection molding, *Materials & Design* 3489–3496 (2009).
12. W. Gacitua, D. Bahr, and M. Wolcott, Damage of the cell wall during extrusion and injection molding of wood plastic composites, *Composites Part A: Applied Science and Manufacturing* 1454–1460 (2010).
13. T. Jiang and G. Zeng, An online extrusion-compression molding method to produce wood plastic composite packaging boxes, *Fibers and Polymers* 804–810 (2019).
14. Y. Zhang, J. Fang, J. Li, Y. Guo, and Q. Wang, The effect of carbon nanotubes on the mechanical properties of wood plastic composites by selective laser sintering, *Polymers* 728 (2017).
15. S. Kain, J.V. Ecker, A. Haider, M. Musso, and A. Petutschnigg, Effects of the infill pattern on mechanical properties of fused layer modeling (FLM) 3D printed wood/polylactic acid (PLA) composites, *European Journal of Wood and Wood Products* 65–74 (2020).
16. A.A.P. Imken, C. Brischke, S. Kögel, K.C. Krause, and C. Mai, Resistance of different wood-based materials against mould fungi: A comparison of methods, *European Journal of Wood and Wood Products* 661–671 (2020).
17. H.A. Babikir, M. Abd Elaziz, A.H. Elsheikh, E.A. Showaib, M. Elhadary, D. Wu, and Y. Liu, Noise prediction of axial piston pump based on different valve materials using a modified artificial neural network model, *Alexandria Engineering Journal* 1077–1087 (2019).
18. A.H. Elsheikh, M. Abd Elaziz, B. Ramesh, M. Egiza, and M.A. Al-qaness, Modeling of drilling process of GFRP composite using a hybrid random vector functional link network/parasitism-predation algorithm, *Journal of Materials Research and Technology* 298–311 (2021).
19. Y. Zhu, G.C. Allen, P. Jones, J. Adams, D.I. Gittins, P.J. Heard, and D. Skuse, Dispersion characterisation of $CaCO_3$ particles in PP/$CaCO_3$ composites, *Composites Part A: Applied Science and Manufacturing* 38–43 (2014).
20. M. Kamal, C.S. Sharma, P. Upadhyaya, V. Verma, K.N. Pandey, V. Kumar, and D.D. Agrawal, Calcium carbonate ($CaCO_3$) nanoparticle filled polypropylene: Effect of particle surface treatment on mechanical, thermal, and morphological performance of composites, *Journal of Applied Polymer Science* 2649–2656 (2012).
21. P.P. Das and V. Chaudhary, Moving towards the era of bio fibre based polymer composites, *Cleaner Engineering and Technology* 100182 (2021).
22. S.K. Najafi, Use of recycled plastics in wood plastic composites–A review, *Waste Management* 1898–1905 (2013).

23. A. El-Kassas and A. Elsheikh, A new eco-friendly mechanical technique for production of rice straw fibers for medium density fiberboards manufacturing, *International Journal of Environmental Science and Technology* 979–988 (2021).
24. Y.V. Vazquez and S.E. Barbosa, Process window for direct recycling of acrylonitrile-butadiene-styrene and high-impact polystyrene from electrical and electronic equipment waste, *Waste Management* 403–408 (2017).
25. N. Mousa, E. Galiwango, S. Haris, A.H. Al-Marzouqi, B. Abu-Jdayil, and Y.L. Caires, A new green composite based on plasticized polylactic acid mixed with date palm waste for single-use plastics applications, *Polymers* 574 (2022).
26. A.E. Ongaro, Z. Ndlovu, E. Sollier, C. Otieno, P. Ondoa, A. Street, and M. Kersaudy-Kerhoas, Engineering a sustainable future for point-of-care diagnostics and single-use microfluidic devices, *Lab on a Chip* 3122–3137 (2022).

Index

Acronyms

MMIP
Magnetic molecularly imprinted polymer, 221
MMPP
Multi-material multilayer plastic packaging, 121
MNP
Micro-nanoplastics, 74
MNPL
Micro/nanoplastic, 76
MP
Microplastic, 81
MSW
Municipal solid waste, 73
NP
Nanoplastic, 81
PA
Poly(amide), 158
PAH
Polycyclic aromatic hydrocarbon, 179, 197
PC
Poly(carbonate), 57, 121
PE
Poly(ethylene), 44, 75, 122, 151
PEMWE
Polymer electrolyte membrane water electrolyzer, 255
PET
Poly(ethylene terephthalate), 75, 108, 116, 147
PGA
Poly(glycolic acid), 175
PID
Proportional, integral and derivative, 213
PKS
Palm kernel shell, 80
PLA
Poly(lactic acid), 15, 44, 133, 175, 228, 278
PMMA
Poly(methyl methacrylate), 60, 164
POM
Poly(oxymethylene), 121
PP
Poly(propylene), 44, 75, 122, 147, 276
PPE
Poly(phenylene ether), 121
PPO
Plastic pyrolysis oil, 81

Chemicals

Boldface numbers refer to Figures

General Index

Also of Interest

Check out these other books by the author published by Scrivener Publishing

Future Trends in Modern Plastics
By Johannes Karl Fink
Published 2024. ISBN 978-1-394-23754-8

Food Safety, Plastics and Sustainability: Materials, Chemicals, Recycling and the Circular Economy
By Johannes Karl Fink
Published 2023. ISBN 978-1-394-17456-0

Porous Plastics
By Johannes Karl Fink
Published 2022. ISBN 978-1-119- 89638-8

Contact Lenses
Materials, Chemicals, Methods and Applications
By Johannes Karl Fink
Published 2022. ISBN 978-1-119- 85735-8

Polymers and Additives in Extreme Environments
Application, Properties, and Fabrication
By Johannes Karl Fink
Published 2021. ISBN 978-1-119- 85137-0

Plastics Process Analysis, Instrumentation, and Control
By Johannes Karl Fink
Published 2021. ISBN 978-1-119-79573-5

Flame Retardants
By Johannes Karl Fink
Published 2020. ISBN 978-1-119-75061-1

The Chemistry of Environmental Engineering
By Johannes Karl Fink
Published 2020. ISBN 978-1-119-70774-5

The Chemistry of Bio-Based Polymers, second edition
By Johannes Karl Fink
Published 2019. ISBN 978-1-119-68131-1

Liquid Silicone Rubber
By Johannes Karl Fink
Published 2019. ISBN 978-1-119- 63133-0

Functional Synthetic Polymers
By Johannes Karl Fink
Published 2019. ISBN 978-1-119-59202-0

3D Industrial Printing with Polymers
By Johannes Karl Fink
Published 2018. ISBN 978-1-119-55526-1

Polymer Waste Management
By Johannes Karl Fink
Published 2018. ISBN 978-1-119-53608-6

Materials, Chemicals and Methods for Dental Applications
By Johannes Karl Fink
Published 2018. ISBN 978-1-119-51031-4

Fuel Cells, Solar Panels, and Storage Devices
Materials and Methods
By Johannes Karl Fink
Published 2018. ISBN 978-1-119-48010-5

Chemicals and Methods for Conservation and Restoration
Paintings, Textiles, Fossils, Wood, Stones, Metals, and Glass
By Johannes Karl Fink
Published 2017. ISBN 978-1-119-41824-5

Additives for High Performance Applications
Chemistry and Applications
By Johannes Karl Fink
Published 2017. ISBN 978-1-119-36361-3

Metallized and Magnetic Polymers
By Johannes Karl Fink
Published 2016. ISBN: 978-1-119-24232-1

Marine, Waterborne, and Water-Resistant Polymers
Chemistry and Applications
By Johannes Karl Fink
Published 2015. ISBN 978-1-119-018486-7

The Chemistry of Printing Inks and Their Electronics and Medical Applications
By Johannes Karl Fink
Published 2015. ISBN 978-1-119-04130-6

The Chemistry of Bio-Based Polymers
By Johannes Karl Fink
Published 2014. ISBN 978-1-118-83725-2

Polymeric Sensors and Actuators
By Johannes Karl Fink
Published 2012. ISBN 978-1-118-41408-8

Handbook of Engineering and Specialty Thermoplastics
Part 1: Polyolefins and Styrenics
By Johannes Karl Fink
Published 2010. ISBN 978-0-470-62483-5

Handbook of Engineering and Specialty Thermoplastics
Part 2: Water Soluble Polymers
By Johannes Karl Fink
Published 2011. ISBN 978-1-118-06275-3

A Concise Introduction to Additives for Thermoplastic Polymers
By Johannes Karl Fink
Published 2010. ISBN 978-0-470-60955-2

Printed in the USA/Agawam, MA
September 16, 2026

910661.075